Nutritional Biochemistry Explained

Nutritional Biochemistry Explained

Alexandra Preston

Alexandra Preston
2014

First Printing: 2014

ISBN: 978-0-646-92513-4

Published by Alexandra Preston, who can be contacted at:
ap_elanora@live.com.au

Distributed by Lulu at: www.lulu.com

Ordering Information:
Special discounts are available on quantity purchases by corporations, associations, educators, and others. For details, contact the publisher at the above address.

U.S. trade bookstores and wholesalers: Please contact Alexandra Preston at:
ap_elanora@live.com.au

Dedication

To Hannah George, my best friend of nine years and counting;
and Emma Zamparutti, Ayesha Ahmed, Candice Allum and Scott
King, my best friends from the Endeavour College of Natural
Health.

Contents

Introduction

The purpose of this book is to explain basic nutritional biochemistry for students of complementary and alternative medicine, nursing, dietetics and other fields where the study of nutrition is necessary. It discusses nutrient absorption, cellular respiration and other topics, based on tutoring notes for the Endeavour College of Natural Health's subject Nutritional Biochemistry.

Nutritional Biochemistry Explained

1: Overview of the Macronutrients

The three macronutrients are carbohydrates, lipids (fats) and protein. These are referred to as macronutrients because the body requires them in relatively large amounts, large enough that the recommended daily intake must be stated in grams as opposed to milligrams.

Carbohydrates

Carbohydrates are only ever made of the elements carbon, hydrogen and oxygen. The number of carbon atoms is usually between 3 and 7, and generally the ratio is 2 hydrogen atoms to every 1 carbon and 1 oxygen atom. Monosaccharides, disaccharides, oligosaccharides and polysaccharides are the four types of carbohydrate; the former two being sugars and the latter two being either starches or fibre. Disaccharides are sometimes categorised as oligosaccharides, or oligosaccharides may be considered to be polysaccharides.

The most common monosaccharides are the 6-carbon (often known as hexose) sugars fructose, glucose and galactose. These are referred to as simple sugars, and because they are already only one unit of sugar they cannot and do not need to be digested any further in preparation for absorption. However, disaccharides, which are two sugar units, must be digested; this is done by a group of enzymes known as the disaccharidases. These are found in the microvilli of the intestinal mucosa cells, and include lactase, sucrose, maltase and isomaltase. Lactose intolerance is caused by a deficiency or absence of the enzyme lactase. Polysaccharides, which are made of more than two sugar units, are usually broken down by the enzyme alpha-amylase, one of the glycosidases. It has been given these names because it breaks the alpha-1,4 glycosidic bonds that join the units of sugar; glucosidase and glucoamylase are two other glycosidase enzymes. Glycosidases are hydrolytic enzymes, which means that they break bonds between carbon and another element, such as hydrogen or oxygen, in the presence of water. This process takes place in the mouth and duodenum, as the pH level in the stomach is too low for glycosidases to survive or function. The enzymes alpha-dextrinase and isomaltase are also in the duodenum; their role is to break the alpha-1,6 bonds in oligosaccharides and dextrins.

Obviously, carbohydrates must first be absorbed into the bloodstream before the body's cells can use them. Monosaccharides are usually the only carbohydrates that can be absorbed; however a small concentration of any disaccharides still present may be absorbed too. The monosaccharides glucose and galactose are absorbed into the mucosal cells of the small intestines (enterocytes) by active transport through the sodium/glucose symporter 1 (SGLT 1) carrier. The side of the enterocytes facing the intestinal lumen is also known as the brush border, and the side facing the inside of the microvilli where the blood vessels are to collect nutrients is the basolateral border. Active transport requires the use of energy in the form of ATP in order to take place, and the SGLT1 carrier was given its name because it must be preloaded with a sodium ion for the glucose to attach to it. However, when high concentrations are present, glucose and galactose are absorbed by facilitated diffusion, which requires a transporter but no energy, through the specific glucose transporter type 2 (GLUT 2). Glucose, galactose and fructose then all cross the basolateral border to exit the enterocytes and enter the bloodstream by passing through the GLUT 2 transporter. When insulin and blood sugar levels are high, the number of these transporters is decreased ("downregulated"), unless there is insulin resistance, which leads to the GLUT 2 numbers being maintained. Although fructose enters the bloodstream via the same transporter as glucose and galactose, it is absorbed into the enterocytes by the GLUT 5 transporter, and then has to use the GLUT 2 transporter down a concentration gradient.

Once absorbed into the bloodstream, all three monosaccharides travel to the liver where galactose and fructose are converted to glucose. The glucose that has not been used for energy by hepatocytes (liver cells) then goes into systemic circulation for use by all other cells in the body. In order to tightly control blood glucose levels and to ensure efficient uptake into cells, there exists the glucose transporters (abbreviated to GLUT), which perform facilitated diffusion. These all have a specific binding site for the glucose to attach to; the ability to undergo a conformational change once the glucose is bound, which moves the glucose to the other side of the membrane and releases it; and the ability to independently reverse the conformational change. The 6 types of GLUT transporter are: GLUT-1, located in the red blood cells, blood-brain barrier, placenta and foetal tissue; GLUT-2, in the liver, pancreatic beta-cells, kidneys and small intestines; GLUT-3 in the neurons of the brain; GLUT-4 in the muscle (skeletal and smooth), fat and heart cells; GLUT-5 in the small intestines and GLUT-7 in the endoplasmic reticulum of hepatocytes. Only the GLUT-4 transporters require insulin for the uptake of glucose.

Lipids

Lipids are hydrophobic, meaning they do not mix with water, and they are made of a glycerol group and fatty acids. They can be simple, compound or derived. The simple lipids are free fatty acids, monoacylglycerols (1 fatty acid); diacylglycerols (2); triacylglycerols (3 fatty acids); and waxes, which are esters of fatty acids with "higher" alcohols, such as sterol esters like cholesterol and non-sterol esters like vitamin A. The compound lipids are phospholipids, which are further categorised into phosphatidic acids such as lecithin, plasmalogens and sphingomyelins; glycolipids and lipoproteins. Derived lipids are derivatives of fats, for example sterols and straight-chained alcohols.

Nutritional Biochemistry Explained

Digestion of lipids presents a challenge to the body, as digestive enzymes are hydrophilic and therefore mix with water, unlike the hydrophobic fats. This is solved by the emulsifying action of bile salts, which break down the fat into smaller pieces, creating more surface area so the enzymes (which are esterases) can reach more ester bonds to cleave them and therefore break down the fat faster. In the mouth, lingual lipase is the enzyme responsible for breaking lipid compounds into triacylglycerols. Then, in the small intestine, pancreatic lipase and colipase continue this role. Additionally, phospholipase breaks down lecithin and other phospholipids; cholesterol esterase breaks down cholesterol esters and retinyl esterhydrolase breaks retinyl esters.

The bile salts also assist in fat absorption. After the enzymes have finished their work, the bile salts, monoacylglycerols, free faty acids and cholesterol form micelles, with the hydrophilic bile salts on the outside. Formation of micelles enables the lipids to easily travel through the small intestine, and when these bump against the brush border, the lipids are taken into the enterocytes by diffusion. The bile salts are not absorbed, but instead return to the liver. Once in the enterocytes, triacylglycerols re-form (re-esterify), and so do phosphatidylcholine and cholesteryl esters.

In the bloodstream, the short-chain fatty acids attach to the hydrophilic protein known as albumin for transport to other cells. However, other lipids travel through the blood as part of chylomicrons. These are lipoproteins, and are necessary because the protein components of chylomicrons make them water soluble, whereas lipids on their own are not, so they could not travel through the blood. There are 3 types of chylomicrons: the VLDLs, with higher triacylglycerols (TAGs) and less cholesterol and protein; the LDLs with less TAGS and more protein and cholesterol; and the HDLs, with the lowest level of TAGs, highest protein and intermediate cholesterol content. The HDLs (often known as "good cholesterol") deliver cholesterol back to the liver.

The proteins also stimulate particular enzymatic reactions and allow the chylomicrons to be recognised by cells. Additionally, adipocytes (fat cells) can make triacylglycerols from glucose. Insulin helps by letting the glucose in to these cells; stimulating the enzyme lipoprotein lipase, which breaks fatty acids away from lipoproteins so they can enter the adipocytes and by stopping intracellular lipase from breaking down fat already stored in the adipocytes.

Proteins

Proteins are water-soluble, nitrogen-containing compounds that are made of individual molecules called amino acids joined together. Peptides are small versions of proteins. Dipeptides are made of only two amino acids; oligopeptides are usually 30-50 amino acids in length and polypeptides can be made of up to thousands of them. Amino acids are classified as either "essential" or "non-essential", with "essential" meaning those the body cannot produce itself and "non-essential" meaning those it can make from other amino acids or molecules such as glucose. The essential amino acids are methionine, leucine, lysine, phenylalanine, histidine, threonine, isoleucine, valine and tryptophan; the non-essential amino acids are alanine, asparagine, aspartate, cysteine, glutamate, glycine, proline, glutamine, tyrosine and serine.

Protein digestion occurs in the stomach with hydrochloric acid and pepsin, and in the small intestine with chymotrypsin, trypsin and finally aminopeptidase and dipeptidase. After this, all amino acids travel across the brush border by energy-dependent transporters. The branch chain amino acids (isoleucine, leucine and valine) are absorbed the fastest; neutral amino acids are absorbed faster than acidic ones and essential ones are also absorbed faster. There can be competition between amino acids that share the same transporter too. However, to cross the basolateral border, amino acids travel by diffusion, or by sodium dependent transport if there are lower amounts of amino acids. Some amino acids may be used by the enterocytes for energy and the production of enzymes or nitrogen-containing compounds, but most end up in the bloodstream. Amino acids travel through blood via proteins such as albumin, transthyretin, and retinol binding proteins. They are then stored in all tissues for energy and the production of structural and functional proteins. These are in a constant cycle of protein breakdown (catabolism) and synthesis (anabolism), except in the brain, which resists protein catabolism.

2: Overview of the Vitamins

Vitamins, like the macronutrients discussed in the previous chapter, are necessary for life as the body cannot produce them on its own. The water-soluble vitamins are the B vitamins along with vitamin C, while the fat-soluble vitamins are A, D, E and K.

Vitamin C

The absorption of vitamin C, also known as ascorbate, in the small intestines varies with intake. At high concentrations, only around 16% is absorbed, but when it is at low levels the absorption rate is about 98%; on average 70-95% is absorbed into the bloodstream. Some of the unabsorbed vitamin C may be used by the intestinal flora. Large amounts of iron can oxidise the vitamin C before it is absorbed, and pectin (14+grams/day) and zinc (9.2+grams/day) speed the loss of vitamin C from urine. It is taken into the enterocytes by the Sodium Dependent Vitamin C Transporter, which requires energy, but if it is oxidised (then referred to as dehydroascorbate), it is absorbed by simple diffusion or the GLUT 1 and 3 transporters. Anion channels move it across the basolateral border and into the bloodstream.

Once in the blood, vitamin C travels freely. If it is in the dehydroascorbate form, glutathione and the enzyme dehydroascorbate reductase can convert it back to ascorbate, thereby restoring its antioxidant properties. The uptake of vitamin C into most cells involves a sodium-dependent carrier, but some, such as leucocytes, use active transport. Vitamin C is found in all cells, with the highest concentrations being in the adrenal and pituitary glands. One function of vitamin C is its responsibility for the function of the enzyme hydroxylase. This enzyme is involved in energy and collagen production, the latter also requiring iron, proline, lysine and glycine. Hydroxylase, with vitamin C and iron, converts proline and lysine to hydroxyproline and hydroxylysine respectively; together these two make up two thirds of all collagen fibres.

Vitamin B1

Vitamin B1, or thiamin, exists in several forms: thiamin mono-, di-, tri- and pyrophosphate, all four of which are found in meats; the non-phosphorylated free form in legumes, whole grains, yeast and wheat germ; thiamin hydrochloride and thiamin mononitrate salt, both of which are only found in commercial supplements. Thiaminases in raw fish catalyse the breakdown of vitamin B1; however, cooking destroys these enzymes. Tannic and caffeic acid also destroy it, but vitamin C and citric acid prevent this; the divalent minerals- those with a +2 ionic charge- inhibit absorption too.

Most thiamin is absorbed in the jejunum, but because only the free form can be absorbed, phosphatases must first cleave any phosphate bonds. Thiamin is often absorbed across the brush border by the thiamin transporters ThTr1 and ThTr2, which require energy, but if levels are over 2.5mg then passive diffusion is used. It may be phosphorylated into a phosphate ester in the enterocyte. ThTr1 is used again to transfer thiamin over the basolateral border, but alcohol (ethanol) blocks absorption into the blood.

Unlike vitamin C, 90% of vitamin B1 travels through blood in erythrocytes (red blood cells) in the form of thiamin diphosphate (TDP). It also travels in its free form and attached to albumin or as thiamin monophosphate (TMP). TMP is converted to TDP (80%) and thiamin pyrophosphate (TPP) (10%) in the liver by the enzyme thiamin pyrophosphokinase, which removes phosphate groups from ATP to do so, this losing energy. Thiamin needs an energy- and sodium-dependent carrier to move into cells, and is found in all cell types, especially in the muscles, liver, kidneys and brain.

Vitamin B2

Vitamin B2 is also known as riboflavin, and is found in a wide variety of foods, especially dairy products, green vegetables, meat and eggs. Riboflavin can be free, protein- or phosphorous-bound, or as the coenzyme derivatives flavin mononucleotide (FMN) and flavin adenine dinucleotide (FAD). It may be destroyed if exposed to sunlight.

In the stomach, hydrochloric acid frees protein-bound riboflavin. After this, in the small intestines, nucleotide diphosphatase and alkaline phosphatase free phosphate-bound riboflavin, while FAD pyrophosphatase and FMN phosphatase convert FAD and FMN to free riboflavin respectively. FAD that is bound to cysteine or histidine cannot be broken down and is excreted unchanged.

Riboflavin is mostly absorbed in the duodenum. Animal sources are usually taken in more easily, with bile being another factor that improves uptake into the enterocytes and blood. However, copper, zinc and manganese bind to riboflavin and FMN, thereby preventing absorption, and alcohol inhibits both its digestion and absorption. At high concentrations, riboflavin moves into the enterocytes by passive diffusion, but normally it requires active transport by riboflavin transporter 2 (RFT2). Once in the enterocytes, it is converted to FMN by the enzyme flavokinase, which uses a phosphate group from ATP. The FMN then crosses the basolateral border by active transport, and travels through the blood bound to plasma proteins, which are albumin, fibrinogens and proteins in the category of globulins.

Riboflavin has many uses, with FAD being the predominant form in tissues. The liver converts FMN to FAD using FAD synthetase; this process is upregulated by thyroid hormones, aldosterone and adrenocorticotrophic hormone (ACTH). Immunoglobulins use riboflavin to produce hydrogen peroxide (H_2O_2), which is necessary for destroying pathogens. Other uses include coenzyme reactions in energy production, vitamin B6 metabolism and the synthesis of vitamin B3 from tryptophan. It is usually taken into cells via active transport by riboflavin-binding proteins, but in the brain only FAD is transported into cells with a high affinity. Riboflavin is found in all cells, with the highest concentrations being in the liver, kidneys and heart where it is stored.

Vitamin B3

Vitamin B3 is usually known as niacin, and sometimes as nicotinic acid or nicotinamide. The main sources of niacin are meat, whole grains, legumes and seeds, but it can be synthesised in the liver from tryptophan; this reaction also requires B2, B6 and iron. It comes in several forms: free nicotinamide in meat; nicotinic acid in plant sources; the nucleotides nicotinamide adenine dinucleotide (NAD) and nicotinamide adenine dinucleotide phosphate (NADP); and bound to carbohydrates (10% absorption) or small peptides called niacinogens.

Nicotinamide and nicotinic acid can be absorbed in the stomach, but this process is easier for the small intestines. Like B1 and B2, alcohol inhibits absorption. If the concentration is less than 3 grams, then B3 is taken in via diffusion through sodium-dependent carriers; if the amount is 3-4 grams then it is absorbed by passive diffusion. In the intestinal lumen and enterocytes, NAD and NADP are converted to nicotinamide by the enzyme glycohydrolase.

In the bloodstream, 15-30% of B3 is found as nicotinamide, and the rest is nicotinic acid, which is bound to plasma proteins. It is taken into cells by passive diffusion, except in the kidneys and erythrocytes, where it requires a carrier, and in the brain where active transport is necessary. Nicotinamide, excess niacin and excess tryptophan are all converted to NAD, which the mitochondria synthesises for energy production.

Vitamin B5

This is also referred to as pantothenic acid, and is found in nearly all foods, especially meats, mushrooms, eggs, legumes, broccoli and avocadoes. 85% of pantothenic acid exists as a component of coenzyme A, and in supplements it is in the form of calcium pantothenate.

During digestion, coenzyme A is hydrolysed to release free pantothenic acid, which is then mostly absorbed in the jejunum. Vitamin B5 is another vitamin that alcohol prevents from being absorbed. At high concentrations, only passive diffusion is needed, but at lower concentrations use of the energy and sodium-dependent multivitamin transporter (SMVT) is necessary. Around 50% of B5 is absorbed, but this drops to 10% when levels approach 10 times the recommended intake. It also crosses the basolateral border by passive diffusion. Most B5 is carried through the blood by erythrocytes, with the rest travelling freely. For uptake into cells, the liver, heart, brain and muscles use the SMVT, but renal cells, adipocytes (fat cells) and cells in the central nervous system only need facilitated diffusion. Pantothenic acid is found in all cells, because its main purpose is the synthesis of coenzyme A, which is used in energy production.

Vitamin B6

Vitamin B6 is also called pyridoxine, and its sources are beef, chicken, salmon, whole grains, walnuts, bananas, navy beans and vegetables. Pyridoxine is the most stable form of B6, and is mostly found in plant sources. The other forms of B6 are pyridoxal (PL); pyridoxamine (PM); pyridoxine phosphate (PNP); pyridoxal phosphate (PLP); pyridoxamine phosphate (PMP) and pyridoxine beta-glycoside, which may not be digestible in mammals. In supplements, B6 is found as pyridoxine hydrochloride and pyridoxine-5-phosphate, the active form. However, processing, which means heating, freezing, canning, prolonged storage, sterilisation and milling of wheat reduces the amount of B6 in foods. Alcohol is also an inhibitor of B6 absorption.

At the brush border, the phosphorylated forms of B6 (PNP, PLP and PMP) have their phosphate groups removed by hydrolysis. This is performed by phosphatase enzymes such as alkaline phosphatase, which is zinc-dependent. Absorption of PN, PL and PM occurs mostly in the jejunum by passive diffusion. Free forms of B6 are more rapidly absorbed, but in high concentrations B6 can be absorbed without having the phosphate groups removed. It also crosses the basolateral border via passive diffusion. Like all other nutrients, B6 is then transported via the portal vein to the liver. In the liver, the PN, PL and PM are phosphorylated again, and some PNP is converted to PLP by a B2 dependent oxidase enzyme. PM travels free in the blood and is immediately available to tissues, but PL, which converts back into PLP, must be dephosphorylated again to PL before it is absorbed by tissues. This is because PLP must bind to albumin and erythrocytes, making it unavailable. B6 is found in all cells, but 75-80% is stored in the muscles and the liver stores 5-10%. It is absorbed by passive diffusion, and is kept from diffusing out of the cells again by being bound to phosphates.

Vitamin B12

The vast majority of vitamin B12 is found in animal foods, but all of it is produced by bacteria. Any B12 in plants is there because of 'contamination' by manure or nitrogen-fixing bacteria; therefore it is only found in small amounts in a few plant foods such as nori, the black trumpet mushroom, tempeh and sauerkraut (the latter two are fermented). B12 is also known as cobalamin, and different forms include cyanocobalamin (-CN), methylcobalamin (-CH$_3$), hydroxocobalamin (-OH), aquocobalamin (-H$_2$O), nitritocobalamin (-NO$_2$) and 5'deoxyadenosylcobalamin. Cyanocobalamin and hydroxocobalamin are usually the form B12 is in for supplements, and in the body cyanocobalamin is converted to aquo- and hydroxocobalamin.

Absorption of cobalamin requires several steps. First, it is released from the peptide bonds in foods by hydrochloric acid and pepsin. Then, R proteins, also known as cobalophilins or haptocorrins, bind to the freed cobalamin in the stomach, but they can do this even in the mouth before it is detached from the proteins. This may protect the cobalamin from intestinal bacteria. In the small intestines, pancreatic proteases break down the R protein by hydrolysis, and then intrinsic factor (IF) binds to the cobalamin. IF is usually necessary for B12 absorption, and reaches saturation, meaning the levels of IF present cannot take any more B12, at 1.5-2 micrograms per meal. The intrinsic factor will then release the cobalamin once it reaches the ileum and attaches to a specific receptor site known as a cubilin, which may be calcium-dependent. Once it is in the bloodstream, cobalamin binds to transcobalamin I, II or III (TCI, TCII, or TCIII).

Various factors affect vitamin B12 absorption. The overall absorption rate of cobalamin is 50%, but 1-3% can be taken in by passive diffusion if at a dose of 1000-2000 micrograms, which is only possible through supplementation. Calcium is necessary for B12 absorption, and 500mg or over of vitamin C may destroy or reduce the B12 if they are ingested less than an hour apart. Cobalamin/B12 absorption is quite slow, with peak blood levels taking 8-12 hours to be reached, due to a 3-4 hour delay between it attaching to the specific receptors and passing through to the circulation. As cobalamin can be recycled, every day 3-8 micrograms is secreted into bile and intestinal secretions, only for it to be picked up by IF and reabsorbed. Both absorption of new B12 and recycling can be affected by conditions such as Crohn's disease which lead to malabsorption.

Unlike the other water-soluble vitamins, B12 can be stored for years. It is taken into cells via energy-dependent (active) transport, by TCII-activated receptors. B12 is mostly stored as adenosylcobalamin in the liver, but is found in all other cells too. Hydroxocobalamin and methylcobalamin make up a minority of stored B12.

Vitamin B9

Vitamin B9 is commonly known as folate, or folic acid. Sources of folate include mushrooms, green vegetables such as spinach and broccoli, peanuts and other legumes, citrus fruits and liver. Raw foods are usually higher in folate because it is destroyed by heat. There are over 150 forms of folate, and all must contain pteridine, PABA and glutamic acid for vitamin activity. Pteroylpolyglutamates are the main forms in food, which contain up to 9 glutamic acid residues; however pteroylmonoglutamates, which only have 1, are the supplemental forms because they are the most stable.

Most folate absorption occurs in the jejunum of the small intestines. First, the polyglutamate forms must be hydrolysed into monoglutamates (MGM folate), otherwise they cannot be absorbed into the bloodstream. This is performed by zinc dependent enzymes known as conjugases (or pteropolyglutamate hydrolases), which are found in bile, pancreatic enzymes and on the jejunal brush border. The proton coupled folate transporters (PCFT) then takes the folate across the brush border; once inside the enterocyte, some of it is attached to folate binding proteins (FBP's). FBP's are used in active transport of folate across the enterocyte, and are already attached to the folate in milk; this complex is more easily absorbed in the ileum. Around half of dietary folate is absorbed due to conjugase inhibitors in some foods, whereas supplemental folate is absorbed more efficiently and via passive diffusion. Conjugase inhibitors are found in lentils, legumes, cabbage and oranges. In a way, alcohol and zinc are also conjugase inhibitors, because they reduce activity of this enzyme too.

In the enterocytes, much of the MGM folate is converted to dihydrofolate, then THF and then 5-methyl THF, with the first 2 reactions involving NAD. MGM folate, DHF, 5-methyl THF and folate bound to FBP's cross the basolateral border to enter the blood vessels via active transport; this is how the folate will later enter the tissues too. All folate in the blood is in the monoglutamate form, and two thirds of this is bound to FBP's. Next, they enter the liver where all are converted to THF, 5-methyl THF or 5- or 10-formyl THF, which can trap folate inside cells because the latter 3 are polyglutamate forms. 5-methyl THF as well as 5- and 10-formyl THF are also the basis of various functional coenzymes. The liver stores half of the body's folate, but it is found in the cytosol and mitochondria of all cells.

Vitamin B7

Vitamin B7 is named biotin, and is widely distributed in foods, especially liver, egg yolks, legumes, soybeans, cereals and nuts. It is also produced by bacteria in the colon, but this form is not useful to humans. Alcohol inhibits biotin absorption, as well as avidin, a protein found in raw egg whites, but heat denatures it so cooked eggs still contain available biotin.

Before it is absorbed in the jejunum and ileum, biotin must be freed from proteins in food by the protease enzymes. It is then absorbed into the enterocytes by the SMVC's, the sodium-dependent multivitamin carriers, which also transport vitamin B5 and lipoic acid. However, large doses only achievable by supplementation can be absorbed by passive diffusion. The SMVC's also carry biotin across the basolateral border, but this time they are not dependent on sodium. Most biotin travels freely through the blood, but some attaches to the plasma protein albumin. Biotin is necessary for some enzymatic reactions involved in the Kreb's cycle, which produces most of the cell's energy, and in gluconeogenesis, which re-forms glucose from other substances, in this case, pyruvate.

Vitamin A

Vitamin A is a fat soluble vitamin mostly found in animal products such as liver, cheese, butter, fish and milk. The term 'vitamin A' refers to retinol, the alcohol form; other forms are retinal (the aldehyde form), retinoic acid (carboxylic acid) and retinyl ester. There are also the pro-vitamin A carotenoids, which are precursors to vitamin A found in plants. Less than 10% of the over 600 carotenoids can be converted to vitamin A, with beta-carotene having the highest conversion rate at 50%. Others include lutein, canthaxanthin, zeaxanthin and lycopene, the red pigment in tomatoes and watermelon. Red, orange and yellow fruits and vegetables contain the highest levels of carotenoids, including carrots, pumpkin, squash and rockmelon. Green vegetables contain some carotenoids, but their pigments are obscured by chlorophyll.

Once they are cleaved from the proteins and fatty acids in food, the retinol and carotenoids are incorporated into micelles. This is because their fat soluble properties make 'hiding' in micelles necessary for them to be able to travel through the small intestines and be taken over the brush border. Fat and vitamin E assist in absorption, but fibre and very large doses of vitamin E inhibit it. In the enterocytes, 1 molecule of beta-carotene turns to two molecules of retinal. One of these converts to retinoic acid, crosses the basolateral border and binds to albumin for transport to the liver, while the other joins with cellular retinoid binding protein II (CRBPII), then NADH+H gives up 2 hydrogen atoms to turn the retinal into retinol. After this, all CRBPII and retinol, whether the retinol was converted from beta-carotene or was originally retinol, joins with fatty acids (palmitic acid) to form CRBPII-retinyl-palmitate. Finally, this joins with the unaltered carotenoids, phospholipids, triacylglycerol and cholesterol to form chylomicrons so they can travel through the lymphatic system. The chylomicrons then deliver the carotenoids and retinyl esters to all cells in the body, with the leftovers being taken to the liver by chylomicron remnants. Some retinol may be metabolised by the liver, some may be re-esterified and stored in the stellate and parenchymal cells of the liver. In order to free stored retinol for transport and use in other tissues, the parenchymal cells must secrete retinol-binding protein (RBP), which requires zinc and dietary protein. One retinol and one RBP join to form holo-RBP; in the plasma this reacts with thyroxine (thyroid hormone) and transthyretin, forming a complex with a half-life of around 11 hours. This enables cells to recognise and absorb the retinol. Carotenoids are transported through the blood in lipoproteins (VLDL, LDL and HDL) for usage by cells, and some is converted to retinol. Retinoic acid is distributed within cells by Cellular Retinoic Acid Binding Protein (CRABP) between the cytoplasm and nucleus in order to control the concentration and make stored retinoic acid free.

Vitamin D

Vitamin D has 2 forms: vitamin D2 (ergocalciferol), found in plants, and D3 (cholecalciferol), found in animal sources such as fish, liver, eggs and butter.

Cholecalciferol/D3 can also be produced by the skin in humans. First, cholesterol is converted to 7-dehydrocholesterol, which is a steroid. Exposure to ultraviolet light from the sun turns this to previtamin D3 (precalciferol), which lasts for 2-3 days before becoming cholecalciferol. This is reabsorbed into the blood by the plasma proteins known as alpha-2 globulins and by D-binding protein (DBP). Reduced sun exposure and low parathyroid hormone affect vitamin D production.

Vitamin D from food sources is absorbed into the enterocytes as part of a micelle, and then into the lymph as part of a chylomicron; this is because it is fat soluble. Half of dietary vitamin D is absorbed, with most being taken in by the ileum despite absorption being fastest in the duodenum. Some is transferred from chylomicrons to DBP once in circulation, which carries 60% of the total circulating vitamin D while chylomicrons carry 40%. Most stored vitamin D is found in the muscles and fat.

In the liver, vitamin D (as cholecalciferol/D3) is converted to 25-hydroxyl cholecalciferol by the NAD-dependent enzyme 25-hydroxylase; after this it travels to the kidney. Low blood calcium stimulates parathyroid hormone (PTH) release, which converts 25-OH-D3 to $1,25(OH)_2$ cholecalciferol and $24,25(OH)_2$ calcitritol. $1,25(OH)_2$-D3 promotes bone mineralisation, while calcitriol, the active form, increases calcium absorption in the intestines and reabsorption in the kidneys. High phosphorous intake decreases calcitriol production, and the enzyme responsible for calcitriol production requires vitamin B2.

Vitamin E

Vitamin E comes in 2 categories: the tocopherols (alpha, beta, gamma and delta) and the tocotrienols (alpha, beta, gamma and delta). Alpha-tocopherol is the most active, and delta-tocopherol is the least active. Both tocopherols and tocotrienols are absorbed across the brush border as part of micelles, and travel across the basolateral border into the lymph inside chylomicrons. The absorption rate can be anywhere from as low as 10% to as high as 80%, with the rate decreasing as the dose increases. Once they arrive in the liver, tocopherols are transferred into HDLs, LDLs and VLDLs by alpha-tocopherol transfer protein (alpha-TTP). Vitamin E may prevent oxidation of LDLs, which is a risk factor for cardiovascular disease. When it is delivered to tissues, it is primarily found in the cell membranes, possibly serving a protective purpose there too.

Vitamin K

There are 3 forms of vitamin K: phylloquinone (K1) from plants, menaquinone (K2) from bacteria and the synthetic supplemental form menadione (K3). Sources of vitamin K include meat, milk, eggs, broccoli, asparagus, kale, cabbage, lettuce, cauliflower, turnips, mustard greens and corn. Colon bacteria also produce menaquinone. Vitamin K absorption is assisted by bile salts and pancreatic juices, and the absorption rate is 40-80%, unless there is fat malabsorption, which leads to a rate of 20-30%. Warfarin is also an antagonist of vitamin K. Phylloquinone is absorbed in the jejunum as part of micelles, and menaquinone and menadione are absorbed in the ileum and colon by passive diffusion. All 3 forms travel across the basolateral border and through the lymphatic system inside chylomicrons, and then they are either delivered to tissues or to the liver where they are transferred to VLDLs. Vitamin K is stored in the liver, heart, bone marrow, salivary glands and brain, and is necessary for blood clotting.

3: Overview of the Minerals

Minerals, akin to the macronutrients and vitamins, are necessary for various functions of the body, serving both structural and functional purposes. Macrominerals (calcium, magnesium, phosphorous, sodium and potassium) are required in large amounts, while microminerals are needed in smaller doses, often in the range of micrograms.

Calcium (Ca)

Calcium is the most abundant macromineral in the body, making up around 1.5% of a person's body weight in an average quantity of 1-1.2 kilograms. 99% of calcium is stored in the bones and teeth, and the other 1% is found in the intra- and extracellular fluids. In supplements, it can be bound to carbonate, acetate, lactate, citrate and malate, gluconate, just citrate, phosphate and orotate. Calcium citrate, citrate-malate, lactate and orotate are the most well absorbed forms. Caution should be observed with calcium carbonate supplements, as there is a risk of aluminium and lead toxicity with those made from dolomite, oyster shells or bone meal.

Absorption of calcium requires solubilisation by stomach acid, calbindin or calcium transporter 1, which performs active transport, and vitamin D as calcitriol. Calcitriol controls brush border entry, movement within the enterocytes and the transport protein shuttles in the basolateral membrane. Calcium absorption is higher when the dose is less than 400 milligrams. However, calcium can be absorbed through the spaces between enterocytes (paracellular absorption) if the dosage is high. If the calcium is bound to fibre, phytates, oxalates or fats then it cannot be taken in, but colonic bacteria may release any calcium that is bound to pectin, a soluble fibre. In the blood, half of all calcium travels freely, 40% is bound to albumin or pre-albumin and 10% forms a complex with citrate, phosphate or sulphate. The absorption rate of calcium is around 30% for adults, and 60% for children. Growth, pregnancy and lactation increase absorption, and so does oestrogen because of its influence on calcitritol. Taking calcium with or as part of food, sugar, protein, low intake and lactose are other factors that increase calcium absorption. One factor that reduces calcium absorption is fat malabsorption, because fats in high enough quantities will bind with calcium and prevent enterocytes from taking it in. Fibre, phytates (possibly) and oxalates also bind with calcium in the intestines and produce the same results. Additionally, caffeine inhibits the enterocyte's ability to hold calcium; ions with a 2+ charge such as magnesium (divalent cations) compete with calcium for absorption; low parathyroid hormone and/or vitamin D3 reduces absorption and so do proton pump inhibitors, as they lead to insufficient gastric acid for calcium solubilisation. However, phytates may have more benefits than drawbacks, as they are associated with a lower colon pH and therefore higher mineral absorption, and may assist in cancer prevention.

When blood levels of calcium are low, production of parathyroid hormone is increased, which is known to leach calcium out of the bones. Another function of this hormone is hydroxylating vitamin D3 in the kidneys a second time to form calcitriol (1,25-OH-D3). This goes to the small intestines to increase calcium absorption, and some remains in the kidneys to prevent the excretion of calcium. Parathyroid hormone production is also stimulated by high phosphorous, which is why a 1:1 ratio of calcium and phosphorous is recommended.

Magnesium (Mg)

Magnesium is the fourth most abundant mineral in the body, but is only second to potassium intracellularly. 55-60% is stored in bones, 20-25% is in the muscles, and the rest of it is found in soft tissue. Some food sources of magnesium are legumes, nuts, oats, seafood, brown rice, chocolate, green vegetables, black strap molasses and carrots. Magnesium can be bound to sulphate (Epsom salts, which can only be absorbed through skin), oxide, chloride, lactate (which is absorbed well), acetate, gluconate and citrate. The absorption rate of magnesium is 30-60%, with lower doses having a higher rate. Like calcium, non-fermentable fibre, phytates, fat malabsorption and other divalent cations such as iron decrease uptake of magnesium. In addition to this phosphorous in high amounts can bind to magnesium and prevent its absorption. Also like calcium, vitamin D, fructose and carbohydrates all increase magnesium uptake.

Magnesium is carried across the brush and basolateral borders via active transport, but it can be taken in by paracellular diffusion if the dose is high enough. Around 55% of magnesium in blood travels freely, about a third is bound to plasma proteins and around 13% is complexed with substances such as citrate or sulphate. Control of blood magnesium levels is reliant on absorption, renal excretion and ion transport, as opposed to hormonal regulation.

Phosphorous (P)

Phosphorous is the second most abundant mineral inside the human body, with 85% being stored in the bones, 14% in the muscles and soft tissue and 1% being used in blood and body fluids.

In food, phosphorous is organically-bound and must be freed by hydrolysis for absorption. This process is zinc-dependent, and is performed by enzymes such as alkaline phosphatase. Calcitriol stimulates both alkaline phosphatase and the production of sodium/phosphorous co-transporters in the brush border.

Phosphorous, now in the form of phosphate (PO_4), travels through the brush border via active transport by sodium-dependent carriers, or by simple diffusion at high concentrations. It then passes through the basolateral border by facilitated diffusion.

However, magnesium, calcium and aluminium inhibit absorption. The majority of phosphate in the bloodstream is in an organic form, bound to phospholipids. These are usually found forming the cell membrane.

Potassium (K)

Potassium is the number one intracellular cation (positive ion), with 95-98% of it being found inside the cells. Its functions are the decrease of blood pressure and calcium excretion; influencing muscle contraction; regulating nerve tissue excitability and assisting in the maintenance of electrolyte balance and pH levels. Over 90% of potassium is absorbed; but the exact sites are unknown. Energy-dependent proton (hydrogen ion)/chloride channels and passive diffusion carry potassium over the brush border, and then potassium channels are used to absorb it into the blood.

Sodium (Na)

This is the major extracellular cation, a counterpart to potassium. It is absorbed across the brush border by sodium/glucose transporters, sodium/proton pumps and sodium channels; then, energy dependent sodium/potassium pumps shuttle it into the blood, where it travels freely. Almost a third of the body's 105 grams of sodium is on the surface of bone crystals, and the rest is in extracellular fluid, nerves and muscle tissue. Excess sodium is associated with hypertension and reduced calcium absorption, but it is necessary for nerve impulses and muscle contraction.

Chromium (Cr)

Chromium is a micromineral found in foods such as meat, organ meats, cheese, brewer's yeast, grains, mushrooms, molasses and brown sugar. Brewer's yeast also contains Glucose Tolerance

Factor (GTF or chromodulin), which is an oligopeptide with 4 chromic ions. Refining foods will decrease chromium levels. Chromium can have various charges, or oxidation states, from 0 to +6. Cr^{3+}, or trivalent chromium, is the most stable, occurs naturally and is an essential nutrient; while Cr^{6+}, or hexavalent chromium, is usually produced by man-made reactions and is a carcinogen, skin irritant, renal and respiratory toxin.

In acidic conditions such as inside the stomach, Cr^{3+} becomes soluble and attaches to ligands such as nitrogen, oxygen or sulfur. Some Cr^{6+} is reduced to Cr^{3+}, thereby reducing the absorbed dose. The chromium then crosses the brush and basolateral borders via diffusion or carrier-mediated transport, and once in the blood competes with iron, copper, cadmium and manganese for binding to transferrin. Chromium will bind to albumin, globulins or lipoproteins if there is no transferrin available. The body stores an average of 4.6 milligrams of chromium, mostly in the kidneys, muscle, heart, bone, liver, blood, spleen and pancreas. It is theorised that chromium may be stored with iron, because they are both transported by transferrin. Cellular uptake of chromium is mediated by receptors. As for absorption from the intestines into the blood, the amino acids methionine and histidine keep chromium soluble in the alkaline pH of the small intestine. Additionally, piccolinate is another stable ligand, and vitamin C is known to improve absorption. Antacids reduce availability of chromium because it reacts with the hydroxyl ions, and so does phytates in large amounts due to their ability to bind to minerals.

Copper (Cu)

Copper is another micromineral, and is found in a wide variety of foods, especially shellfish, nuts and seeds. It has a cuprous (Cu^{1+}) and cupric (Cu^{2+}) form.

In the small intestines, the amino acids bound to copper are cleaved from it by proteases. Then, the free copper ions travel through the brush border by either Ctr1 (copper transporter 1) or DMT1 (divalent metal transporter 1). Next, the copper either joins to amino acids or glutathione, or is taken by a 'chaperone' to the basolateral border, where it crosses into the bloodstream by active transport and attaches to albumin or alpha-2 macroglobulin. However, some copper stays in the enterocytes to be stored in metallothionein, or form superoxide dismutase (SOD) or cytochrome C oxidase. Liver cells also use copper to produce metallothionein, and all cells require it as a co-factor for various enzymes. Factors that reduce copper absorption are antacids because they raise stomach pH; zinc, which inhibits copper storage in enterocytes; phytates and molybdenum because they bind with copper and therefore prevent absorption; as well as vitamin C, iron, calcium and phosphorous, which also reduce availability. Amino acids, especially histidine, methionine and cysteine; glutathione; and organic acids such as citric, lactic and gluconic acids all improve absorption. Citrate (citric acid) in particular is effective in this by forming a stable complex with copper.

Iodine (I)

Iodine is found in most foods, but levels are often a reflection of the soil content so the best sources are seafood (including seaweed). It is usually in its ionic form, which has a single negative charge (I^-). Goitrogens, or isothiocyanates, inhibit iodine's availability by competing with uptake into the thyroid. These are found in the cruciferous vegetable (*Brassicaceae*) family, and include broccoli, kale, cauliflower and cabbage.

As for absorption, iodine is taken into the salivary glands, stomach and small intestines via passive diffusion. Any iodine in the form of iodate (I^{3-}) is oxidized to iodine in a glutathione-dependent reaction. Tyrosine and other amino acids also assist in iodine uptake. After this it travels through the blood where the thyroid gland will absorb 70-80% of it to make the thyroid hormones T3 and T4, via sodium-dependent active transport.

Iron (Fe)

Iron comes in both haeme and non-haeme sources, and can be ferric (Fe^{3+}), or ferrous (Fe^{2+}). All meats contain haeme iron, while plants such as nuts and green vegetables have non-haeme iron.

In humans, 65% of iron is used to make haemoglobin, 10% is used in myoglobin and 1-5% is found as parts of enzymes. Polyphenols in coffee and tea can reduce iron absorption by up to 40%, while oxalic acid from spinach, chard, berries and chocolate can inhibit it up to 60%. Other divalent metals and phytates also reduce the availability of iron. Factors that improve iron absorption are acids, including vitamin C and citric acid; sugars and meats. Iron is absorbed over the brush border by DMT1, after it is reduced (an electron is added) from Fe^{3+} to Fe^{2+} if it is from a plant source. Haeme from animal protein is taken in by haeme carrier protein (HCP1), and then its components are separated into iron and protoporphyrin by haeme oxygenase. The free iron is taken into the basolateral border's ferroportin transporter by either mobilferrin or the amino acids cysteine and histidine, where it attaches to transferrin or hephaestin for transport in the blood. However, some iron stays in the enterocytes, and is stored in ferritin or hemosiderin. Cells absorb iron by performing endocytosis to take in the transferrin, and then the iron is separated from the transferrin for immediate use by the cell or to be stored in ferritin or hemosiderin. Iron is easily freed from storage because ferritin is constantly being broken down and reformed.

Manganese (Mn)

Manganese is found in many foods, such as bread, legumes, root vegetables, nuts and milk; and it can be either divalent (Mn^{2+}) or trivalent (Mn^{3+}). Factors that improve manganese absorption are the ligand histidine citrate and consuming it with or as part of animal products. Iron competes with manganese for absorption, while oxalates, fibre and phosphorous form complexes with it which reduces the amount of available manganese. Manganese may be absorbed into the enterocytes via DMT1, and travels through the blood freely or bound to transferrin, albumin or alpha-2 macroglobulin. It is stored in most cells, particularly the bones, liver, pancreas and kidneys.

Molybdenum (Mo)

Molybdenum is found in many foods, but amounts vary based on soil concentration. In the human body it usually has a +4 or +6 charge (Mo^{4+} and Mo^{6+}) and is typically bound to sulphur or oxygen. Low amounts of molybdenum are absorbed by active transport, but higher doses only require simple diffusion. It is then transported in the blood by either albumin or alpha-2 macroglobulin. Most of the body's molybdenum is stored in the liver, kidneys and bones.

Selenium (Se)

Selenium has more sources than any other trace element, but the main ones are nuts (walnuts, Brazil nuts and cashews), garlic, onions, organ meats, grains and seafood. Mercury in seafood will decrease the bioavailability of selenium because they bind to each other, thereby preventing either from being absorbed. Like other microminerals, the levels in other foods are dependent on the soil concentration. The organic forms of selenium are selenomethionine, selenocysteine, selenocystine and Se-methyl methionine. In these forms, the selenium is replacing the sulphur normally present in these amino acids. The inorganic forms, found in some plant sources, are selenite (SeO_3) and the possibly better absorbed selenate (SeO_4). Selenoaminoacids have an absorption rate of 50-80%, and the average rate for both organic and inorganic forms is 45-70%. Factors that increase selenium absorption are vitamins A, C and E, as well as low glutathione. Also, besides mercury, phytates reduce selenium absorption.

In the small intestines, organic selenium travels over both the brush and basolateral borders by amino acid transporters, and then it travels through the blood freely. Inorganic selenium, however, is transported over the brush border by sodium-dependent active transport if it is selenate or by simple diffusion if it's in the selenite form. It is unknown how they travel over the basolateral border, but they do also travel freely through the blood.

Zinc (Zn)

Zinc is found in meat, including seafood and especially oysters, legumes, grains, milk, eggs and vegetables. It is almost universally found as Zn^{2+}. Zinc from meats is absorbed the most efficiently, but it must be freed from proteins by hydrolysis in the stomach. However, heating reduces zinc availability due to the formation of complexes. Citric and picolinic acid, amino acids, pancreatic juices, glutathione and tripeptides all enhance zinc absorption; factors that inhibit it are phytates, oxalates, polyphenols, fibre, folate, other divalent minerals and antacids.

After it is cleaved from proteins by digestive enzymes in the stomach, zinc travels to the small intestine where it is absorbed by DMT1, ZIP4 or paracellularly if the dose is over 20mg. Some zinc is stored in metallothionein by the enterocytes or used for functional purposes. Most, however, is transported by CRIP or travels on its own to the ZnT transporters in the basolateral border to enter the bloodstream, where it binds to albumin for transport. Zinc is stored in all tissues, but the highest concentrations are in the liver, kidney, muscles and skin.

Boron (B)

Boron is found in a variety of foods, including avocadoes, meats, peanuts, grapes, nuts and vegetables. It can be in the form of either boric acid or sodium borate (borax). The vast majority of boron in food is absorbed, which takes place via passive diffusion.

This is then stored in the bones, teeth, nails and hair, and may play a structural role. The total boron content in the body is between 3 and 20 milligrams.

Vanadium (V)

Vanadium is only found in very low quantities, and is mainly found in seafood, black pepper and grains. It is usually pentavalent (V^{5+}) or divalent (V^{2+}). Vanadium is absorbed poorly (less than 5%), and the highest concentrations of it are in the bones, kidneys, spleen and liver. However, even in these places the level of vanadium is very low.

4: Energy Production and Storage

The fate of monosaccharides, whether they will be used for energy production or stored, is dependent on the body's needs at that time. Hormones such as insulin, glucagon, adrenalin and corticosteroids control the activity of metabolic pathways involved in energy production and storage. Allosteric enzymes also control some of these pathways, because their activities can be altered by negative modulators, which slow reactions, and positive modulators which increase the rate of reactions. Metabolic pathways are grouped into 2 categories: anabolic ("building up") and catabolic ("breaking down"). The anabolic pathways are protein synthesis, glycogenesis, lipogenesis and gluconeogenesis; the catabolic ones are proteolysis, glycogenolysis, lipolysis and glycolysis. "Lysis" means to break something down, and "genesis" means to create.

Glycogenesis

Glycogenesis is the pathway that converts glucose molecules to glycogen for storage. Glycogen is a reserve of instant energy, mainly produced by the liver, and stored in both the liver and muscles. In fact, around 7% of the liver's weight is from glycogen.

During glycogenesis, glucose molecules are let into the liver cells (hepatocytes) by insulin. Then, 1 atom of magnesium and 1 ATP molecule are used to convert it to glucose-6-phosphate (G-6-P). G-6-P has a phosphate group on the 6th carbon. Muscle cells use hexokinase to add this on while hepatocytes use glucokinase, which can be induced by insulin. Next, the G-6-P becomes glucose-1-phosphate as the enzyme phosphoglycomutase transfers the phosphate group from the 6th carbon to the 1st. After this, uridine triphosphate (UTP) gives up two of its phosphates, then this uridine monophosphate attaches to the phosphate in G-1-P to make UDP-glucose. This is necessary to prevent the glucose from crystallising.

Insulin and the phosphorylation of serine residues then stimulate glycogen synthase to attach the UDP-glucose to the glycogen chain, which also requires the protein glycogenin as a primer. As more glucose is added, the growing glycogen chain branches in order to increase solubility, compactness and the number of readily available glucose molecules when energy is needed.

Glycogenolysis

Glycogenolysis is the reverse of glycogenesis; the breaking down of stored glycogen back into glucose. Glycogen phosphorylase helps break glucose molecules off of glycogen, which are in the form of G-1-P. This is regulated by the hormones adrenalin and glucagon (therefore these have an opposite effect of insulin) and by high levels of adenosine monophosphate (AMP). High AMP is a sign that more adenosine triphosphate (ATP) must be produced for energy. ATP is needed to store energy in a way that is accessible to cells, and high levels of this keeps glycogen phosphorylase inactive because it means there is enough energy. After this, the G-1-P turns to G-6-P, and must have its phosphate removed by glucose-6-phosphatase to be able to enter the blood and be available to all cells in the body. Only the liver can manufacture this enzyme, therefore it is the only source of glycogen that can control blood glucose levels, whereas muscle cells only use their glycogen for themselves.

Glycolysis

Glycolysis is the oxidation of glucose, and is an important pathway in the use of sugars to produce cellular energy. Under aerobic conditions (in the presence of oxygen), 2 molecules of ATP are used to convert 1 molecule of glucose to 2 of pyruvate; 4 molecules of ATP and 2 NADH are also produced.

There are 10 reactions in glycolysis. First, hexokinase or glucokinase, 1 ATP and 1 of magnesium convert **glucose** to **G-6-P**. Then, phosphohexose isomerase turns it to **fructose-6-phosphate**.

Thirdly, phosphofructokinase 1, an ATP molecule and a magnesium ion are used to turn this to **fructose 1,6-bisphosphate**, which is split by aldolase to **glyceraldehyde-3-phosphate** and **dihydroxyacetone phosphate** (triosephosphate isomerase turns the latter into **G3P**). These are both converted to **1,3-bisphosphoglycerate** by G3P dehydrogenase, which also turns 2 NAD molecules to NADH+H by adding hydrogens. The first 2 molecules of ATP are produced form ADP when phosphoglycerate kinase and magnesium change them to **3-phosphoglycerate**. These are converted to **2-phosphoglycerate** by phosphoglycerate mutase, then to **phosphoenylpyruvate** by enolase. Finally, pyruvate kinase and 1 magnesium ion change them to **pyruvate**, thereby also producing 2 more ATP molecules. The NADH+H help to generate more ATP in the electron transport chain; therefore the true net yield of ATP from glycolysis is 6-8 molecules per glucose molecule. Pyruvate can also be produced by the amino acids methionine, cysteine, tryptophan, alanine and threonine.

The Kreb's Cycle

The Kreb's Cycle is also known as the Citric Acid or Tricarboxylic Acid cycle (the TCA cycle). Whereas glycolysis occurs in the cytosol, this takes place in the mitochondria, and over 90% of energy production from food is a result of this pathway. The TCA cycle requires thiamin (as TDP), riboflavin (as FAD), vitamin B3 (as NAD), vitamin B5 as coenzyme A, lipoic acid, iron and magnesium.

In the TCA cycle, first of all a transporter protein takes the pyruvate into the mitochondria. Then, it loses a carboxylic group, which becomes carbon dioxide, and is now a 2 carbon molecule. After this, NAD removes 2 of its hydrogens to make NADH+H, and coenzyme A attaches to the remaining molecule to form **acetyl CoA**. Besides pyruvate and coenzyme A, acetyl CoA can be produced from the amino acids isoleucine, leucine, lysine and tryptophan. When acetyl CoA joins the TCA cycle, it combines with **oxaloacetate** (which can be made from aspartate) to form **citrate** (or citric acid). Citrate becomes **isocitrate**, which then loses hydrogen by isocitrate dehydrogenase to turn into **alpha ketoglutarate**. This produces energy/ATP by later reactions in the respiratory chain and by reoxidating NADH+H, but also requires magnesium and manganese. Alpha ketoglutarate can also be formed from arginine, ornithine, hydroxyproline, proline, glutamate and histidine. Next, this is decarboxylated by a dehydrogenase complex to **succinyl CoA**, which can be made from isoleucine, valine and methionine too. Succinyl CoA then has its thioester bond broken, which produces enough energy to add another phosphate to guanosine diphosphate (GDP). The resulting G. triphosphate is now able to donate a phosphate to ADP to make ATP, and the succinyl CoA is now **succinate**. The succinate becomes **fumarate** by dehydrogenation, which also produces $FADH_2$ from FAD. Fumarate sometimes comes from phenylalanine and tyrosine too. After this reaction, fumarase adds a water molecule across fumarate's double bond to form **malate**, which is dehydrogenated by malate dehydrogenase and becomes oxaloacetate, beginning the cycle again. NADH+H is the other product of the "final" reaction. All of the NADH+H and $FADH_2$ molecules then go into the electron transport chain, along with those from glycolysis and the formation of acetyl CoA, to produce the majority of the cell's ATP.

The Electron Transport Chain

The 2 previous pathways produced a total of 8 NADH+H and 2 $FADH_2$ molecules for every molecule of glucose, which then go on to the electron transport chain (ETC). This takes places in the inner mitochondrial membrane. Hydrogen atoms from the aforementioned B2 and B3 coenzymes are sent down a cascade of fat soluble enzyme complexes in order to produce energy as ATP.

These complexes require the nutrients iron, copper, vitamin B2 and coenzyme Q10.

The ETC is made of series of 4 electron transporters that are all embedded in the inner mitochondrial membrane. Electrons and the hydrogen atoms from NADH+H and $FADH_2$ are pumped from one side of the membrane to the other, with the final reaction being the oxygen and hydrogen combined to form water. Without oxygen, the ETC cannot function, and therefore energy production would be impaired to the point that survival is impossible. This is one reason why most life forms need oxygen.

Cori Cycle

During glycolysis, glucose is converted to pyruvate. However, when there is insufficient oxygen, it cannot enter the TCA cycle, and therefore only 2-8 ATP molecules will be produced from 1 unit of glucose. This is another reason why oxygen is vital for survival, as 2-8 ATP units per unit of glucose is severely insufficient for a cell to function. Actually, without oxygen, not even glycolysis could be performed, due to the accumulation of pyruvate, unless the excess pyruvate is transformed into lactic acid. The Cori cycle is the pathway that converts lactic acid back to glucose, and it takes place in the liver. Glutamine speeds up this pathway because it shuttles lactic acid to the liver. Once in the liver, lactate dehydrogenase turns lactic acid back to pyruvate, which then undergoes a series of reactions, similar to glycolysis in reverse, to become glucose once more.

Gluconeogenesis

Gluconeogenesis is the synthesis of glucose from non-carbohydrate substances, such as lactate, pyruvate, glycerol and some amino acids. Fatty acids, glycerol and amino acids are first converted to pyruvate before they become glucose. As stated above, the conversion of pyruvate to glucose is similar to the reverse of glycolysis, but there are a few differences. The number of ATP molecules used up is 6 as opposed to 2, and gluconeogenesis is the only case where phosphoglycerate kinase requires energy. Also, pyruvate turns to oxaloacetate, then phosphoenylpyruvate, and finally 3-phosphoglycerate before the changes it undergoes match glycolysis. Although the body can use fatty acids and ketone bodies for energy, glucose is "preferred" over these substances, so ketogenesis only occurs during starvation or when someone is on a very low carbohydrate diet.

Lipogenesis

Lipogenesis is the process where carbohydrates are converted to fat for storage. This happens when glycogen stores are filled, because when calories are ingested in excess of the body's needs, it must be either used as energy or stored as glycogen or fat. Lipid reserves are in a state of dynamic equilibrium, meaning that, all the time, some fats are being broken down for energy while others are being sythesised and stored. In fact, the average "lifespan" of each lipid molecule is 2-10 days in rats.

The pathway of Lipogenesis takes place in the cytoplasm. It starts with an acetyl CoA molecule, which has other 2 carbon units added to it in order to form a fatty acid, usually the 16 carbon palmitate. Many of the enzymes used to make these fatty acids are part of a multienzyme complex called fatty acid synthase. Acyl carrier protein (ACP) is vital to lipogenesis, and is a rotating protein that the growing fatty acids are attached to. Condensing protein (CE) contains cysteine and aids in the growth of fatty acids. Energy is required, and so are hydrogen atoms because the reactions involved are reductions (the addition of hydrogen or electrons). Also, the fourth reaction involves the loss of a water molecule, and the third causes a loss of a carbon dioxide, as does the 9th/final reaction, which converts butyryl CE-malonyl ACP to palmitate.

Lipolysis

Lipolysis is the catabolism of triglycerides into fatty acids and glycerol by lipases, the opposite of the above pathway. This is under hormonal control; adrenalin and glucagon stimulate lipolysis while insulin inhibits it (the opposite is true for the synthesis of fats from fatty acids). The 3 enzymes that are involved are triacylglycerol lipase, diacylglycerol lipase and monoacylglycerol lipase, but only the first must be activated by adrenalin.

Beta-Oxidation

This is a continuation of the process started by lipolysis; the conversion of fatty acids into acetyl CoA for gluconeogenesis or for direct entry into the Kreb's cycle. Beta-oxidation occurs in the mitochondria, unlike lipogenesis. First, coenzyme A activates the fatty acid by attaching to it, and then carnitine transports them into the mitochondrial matrix. Once it is inside the matrix, the palmitoyl CoA (palmitate + acetyl CoA) is oxidised to lose 2 hydrogens by FAD and acetyl CoA dehydrogenase, and become acyl CoA. The FAD becomes $FADH_2$ and enters the electron transport chain to produce energy. Acyl CoA then gains a water molecule with the enzyme enoyl CoA hydratase to become beta-hydroxyacyl CoA. This is oxidised to lose 2 hydrogens again by its dehydrogenase enzyme and NAD to become beta-ketoacyl CoA and form NADH+H, which can also now enter the ETC. Then, another coenzyme A is added to the beta-ketoacyl CoA by acetyltransferase, thereby turning it into acetyl CoA so it can go to the Kreb's cycle.

Cholesterol

Cholesterol is an extremely important substance for the body, because it is a precursor to glucocorticoid and steroid sex hormones, bile salts and vitamin D, and is also required for structural integrity of cell membranes. Both cholesterol from food and that synthesised by the body is transported by the lipoproteins (HDL, LDL and VLDL), and so are cholesteryl esters, which are the form that cholesterol is stored in cells as. Just under half of the body's cholesterol is from biosynthesis; this occurs in the cytoplasm and microsomes and, like lipogenesis, uses the acetate group from acetyl CoA. The acetyl CoA is joined to acetoacetyl CoA to form HMG-CoA, which HMG-CoA reductase converts to mevalonate. This is the enzyme that is blocked by statin drugs. Next, the mevalonate is changed to isopentenyl pyrophosphate (IPP), and loses a carbon dioxide. After this the IPP is converted to squalene, and finally to cholesterol.

As stated above, there are 3 types of lipoproteins: LDL, HDL and VLDL. Low density lipoproteins (LDL) consist of a cholesterol core (50%) with protein (20%) and phospholipids surrounding it. This is the major carrier for cholesterol made by the body (endogenous), and is responsible for delivery to cells. Cells have LDL receptors on their membranes in order to allow the uptake of cholesterol. High blood levels of LDL are associated with an increased risk of cardiovascular disease, but it is other factors that actually trigger atherosclerosis. High density lipoproteins (HDL) transport excess cholesterol (around 22% of their weight) back to the liver, and proteins make up half of their weight. Regular exercise raises blood levels of these. Their role of removing cholesterol from tissues is thought to be why high levels of HDLs assist in the prevention of cardiovascular disease/atherosclerosis; however they do have other protective properties. HDLs are a major carrier of lipid hydroperoxides, which contribute to oxidative stress in tissues. It takes these offending molecules back to the liver to be removed and away from where they could contribute to atherosclerosis. HDLs also contain the enzymes paraoxonase 1 (PON1), lipoprotein-associated phospholipase A2 (LpPLA2) and LCAT, which break down oxidised phospholipids and therefore prevent more oxidative stress. Additionally, methionine and other

Nutritional Biochemistry Explained

sulphur-containing amino acids in HDLs inactivate phospholipid hydroperoxides. Besides HDL's antioxidant properties, it blocks the release of chemicals that stimulate inflammation and inhibits inappropriate blood clotting. Finally, there are the very low density lipoproteins (VLDLs). These are made of around 75% triglycerides and 25% cholesterol, and are the main lipoprotein responsible for delivering lipids to cells.

Atherogenesis, the formation of atherosclerotic plaques that lead to cardiovascular disease, is specifically caused by macrophages engulfing and modifying *oxidised* LDL molecules. When the macrophages have "eaten" too many, they become foam cells and produce various pro-inflammatory chemicals. These contribute to inappropriate leucocyte accumulation (leading to more unnecessary inflammation), cause proliferation of smooth muscle cells, remodel the extracellular matrix and inhibit collagen production. Apart from these actions, which narrow the arterial lumen by thickening the wall, they also express potent tissue factors that help stimulate the formation of blood clots, which may eventually break off and cause a heart attack or stroke. Therefore, cardiovascular disease is not simply caused by "high cholesterol", but primarily by oxidation and inflammation.

5: Antioxidants & Free Radicals

In chemistry terms, oxidation can mean a molecule has gained oxygen, lost hydrogen or lost electrons, while reduction means a gain of hydrogen, loss of oxygen or gain of electrons. Therefore, an oxidant removes electrons or hydrogen, or adds oxygen, and a reductant removes oxygen, or adds electrons or hydrogen. A free radical is an oxidant, as it is an atom or molecule with one or more unpaired electrons, because each atomic orbital is meant to hold a maximum of 2 electrons. They are unstable, and will steal electrons from molecules in tissues, however important those molecules may be. Free radicals may contain oxygen (Reactive Oxygen Species/ROS), nitrogen (RNS), sulphur in the case of thiyl (RS), or chloride in the case of trichloromethyl (CCl_3).

Reactive Oxygen Species are formed or enter the body upon exposure to substances such as ozone, smog, chemicals, drugs, radiation, high oxygen and during some normal physiological processes. Some ROS are superoxide (O_2^-), hydroxyl ($OH\cdot$), hydroperoxyl, alkoxyl and peroxyl; others include ozone (O_3), singlet oxygen, hypochlorous acid and hydrogen peroxide (H_2O_2), but these four are not free radicals. Free radicals can damage lipids, whether they are in cell membranes or lipoproteins (which contribute to atherosclerosis); they can damage organelles such as mitochondria or DNA, causing mutations; they can degrade proteins and even destroy red blood cells. They also "breed" other free radicals, causing further damage. This damage contributes to aging, cancer, cardiovascular disease, cataracts and complications in diabetes. Antioxidants, on the other hand, are compounds that prevent free radicals from reacting with the tissues. They can donate electrons to free radicals in order to stabilise them, or stop them from causing oxidative damage without electron donation. This leads to benefits such as a reduction in LDL oxidation, which slows LDL uptake into macrophages and thereby prevents them from becoming foam cells. Vitamins A, C and E; the carotenoids; glutathione; coenzyme Q10; lipoic acid and selenium are all antioxidant nutrients.

The body is exposed to and generates a variety of free radicals, or reactive species, every day. Superoxide (O_2^-) is one of these. It may be generated by oxygen reacting with compounds such as catecholamines, adrenalin, folate and tetrahydrofolate; in the ETC from electrons leaking onto oxygen, or by the body intentionally making its own free radicals (auto-oxidation). Electrons reacting with oxygen and CoQH in order to produce ATP also make superoxide. Cytochrome P450 enzymes hydroxylating fat-soluble substrates to make them more polar, by transferring electrons from NADPH and binding molecular oxygen to them, produce superoxide too. Activated leucocytes (white blood cells), such as macrophages and neutrophils create superoxide intentionally to destroy pathogens and attract more leucocytes. This is often referred to as a respiratory burst.

Hydrogen peroxide (H_2O_2) is not technically a free radical, but considered to be a ROS because it can easily cause damage to cells and tissues. It is produced when superoxide dismutase or vitamin C destroy superoxide as an intermediate in the process of removing this free radical. H_2O_2 is also formed by oxidation of compounds in peroxisomes. These are organelles that degrade unwanted molecules such as excess fatty acids. Additionally, an increase in H_2O_2 is caused by injury from ischaemia, due to xanthine oxidase, ETC disruption and neutrophil activation, and by the copper-dependent amine oxidase.

In the presence of iron, because it can donate another electron, hydrogen peroxide and superoxide can react to form the hydroxyl radical ($\cdot$OH), which is called the Haber-Weiss reaction. Exposure to gamma rays splits water to form hydroxyl radicals; this is one reason why radiation is so dangerous. The Fenton reaction also produces the hydroxyl radical, when iron donates an electron that splits peroxide into a hydroxyl group and a hydroxyl radical. The hydroxyl radical is considered to be one of the most dangerous free radicals, able to attack any molecule in the body and thought to play a major role in lipid peroxidation, so removal of this is very important.

Another 2 free radicals are peroxyl (O_2^-) and hydroperoxyl (HO_2^-). These are oxygen centred radicals produced when superoxide reacts with hydrogen and additional electrons, and are worse than superoxide itself. Hydroxyl radicals, besides directly causing damage to cells, can form lipid carbon-centred radicals ($L^.$) by attacking unsaturated fats in cell membranes, LDLs and other places. Oxygen and peroxyl radicals are also able to produce $L^.$, with oxygen having the ability to make hydroxyperoxyls ($HO^._2$) and peroxyl being able to form lipid peroxides (LOOH). LOOH can react with iron to generate other free radicals and initiate chain reactions with other unsaturated fatty acids. Unlike other free radicals, singlet molecular oxygen ($^1O_2$) does not "steal" electrons or hydrogen, but has its peripheral electron excited to the orbital above what it usually occupies. It can be generated during the respiratory burst, lipid peroxidation or by UV radiation.

As stated above, unchecked ROS and RNS cause or contribute to cancer, cardiovascular disease, diabetes complications and other diseases. This is why antioxidants are necessary, as they reduce or eliminate these problematic molecules. The primary mechanism used to remove superoxide radicals is by converting it to other compounds. Vitamin C (ascorbate), aided by zinc, copper and manganese, can donate 2 hydrogen atoms to superoxide, thus forming H_2O_2 from superoxide and dehydroascorbate from ascorbate. The enzyme superoxide dismutase (SOD) is faster at converting superoxides than vitamin C, but also requires zinc, manganese and copper.

Vitamin C can also eliminate hydrogen peroxide, and so can the enzymes glutathione peroxidase, catalase and myelperoxidase. Glutathione peroxidase neutralizes H_2O_2 and other peroxides, and is found in the mitochondria and cytosol. It requires selenium as a co-factor, as well as the amino acids cysteine, glycine and glutamine, because glutathione is made from these three. Glutathione is versatile, able to function in both aqueous and lipophilic areas of the body. After it is oxidised from neutralising the free radicals, glutathione needs to be regenerated by selenium, B2 (as FAD) and B3 (as NADPH+H). Catalase, however, is haeme iron-dependent and mainly found in the peroxisomes and leucocytes. Myelperoxidase is also dependent on haeme iron, and uses H_2O_2 to produce the potent, toxic radical hypochlorous acid (HOCl) for the respiratory burst in activated leucocytes.

Several substances are used by the body to defend itself against hydroxyl radicals. Vitamin C donates 1 hydrogen to neutralise it into water, and works in aqueous regions (such as plasma) only. Glutathione and dihydrolipoic acid can fight hydroxyl radicals in both aqueous and lipophilic (e.g. cell membranes) environments. Oxidised glutathione (GS·) must join in pairs before they are regenerated, while dihydrolipoic acid becomes lipoic acid again. Apart from vitamin C, uric acid, coenzyme Q10 and metallothionein also work against hydroxyls in aqueous areas.

The reduction, or neutralisation, of lipid peroxides as well as carbon-centred, peroxyl (LOO·) and hydroperoxyl radicals requires several antioxidant nutrients. One of these is vitamin E, a fat-soluble vitamin that is located in the cell membranes and lipoproteins. It works by donating a hydrogen and then stabilising the unpaired electrons with its chromanol ring. Thus, vitamin E breaks chain reactions started by neutralising carbon centred and peroxyl radicals so they cannot damage other polyunsaturated fatty acids. Beta-carotene performs a similar function, except it functions inside the cell as opposed to on the surface. Manganese, vitamin C and coenzyme Q10 also reduce peroxyl radicals to lipid peroxides (LOOH), by donating an electron and a hydrogen respectively. This is why increasing intake of coQ10 helps to prevent atherosclerosis.

However, lipid peroxides can still damage cell membranes due to their polarity. Glutathione peroxidase uses glutathione (GSH- its reduced form) to remedy this, by donating 2 hydrogens to convert LOOH into hydroxyl acid (LOH) and a water molecule.

Singlet molecular oxygen ($^1O_2$) can be returned to its ground (non-damaging) state by carotenoids, vitamin C and lipoic acid (a thiol, as it contains sulphur). Carotenoids, especially lycopene and beta-carotene, can neutralise hundreds of these due to their conjugated double bond systems. They work by absorbing the excess energy from $^1O_2$, then releasing it as heat without chemical change, so they do not need regeneration.

Most antioxidants, however, must be regenerated so they can eliminate free radicals multiple times and thereby help prevent disease. This is usually performed by other antioxidant nutrients. Vitamin E is regenerated by vitamin C, coenzyme Q10 and glutathione (GSH) via hydrogen donation in the membrane surface. The compounds responsible for vitamin C regeneration are vitamin B3 (as NADH), dihydrolipoic acid, glutathione and thioredoxin. This is also performed by hydrogen donation. Glutathione, which joins together in pairs when oxidised, is regenerated by glutathione reductase (needs B2 as FADPH+H) and dihydrolipoic acid by having its hydrogen returned. Coenzyme Q10 is regenerated from ubisemiquinone (its oxidised form) back to ubiquinol by dihydrolipoic acid and thioredoxin, which needs FAD and selenium. Thioredoxin in turn is regenerated by thioredoxin reductase and NADPH+H.

Despite the antiaging and disease preventing effects of antioxidants being proven in *in vitro* and non-experimental studies, experimental studies of singular antioxidants often do not support these findings. This is because antioxidant function involves a complex web of interdependence as opposed to any of them being able to work effectively in isolation.

6: The Formation of Blood Cells

The formation of blood cells is also known as haematopoiesis. Blood cells have short lifespans, and so they must constantly be replaced in matching numbers.

First, erythropoietin is secreted by the kidneys, which stimulates the bone marrow to grow more lymphoid and myeloid stem cells. These are self-renewing and usually safe from harm. Lymphoid stem cells eventually become lymphocytes, which are T cells and B cells. Myeloid stem cells become platelets, neutrophils, macrophages, basophils, eosinophils and erythrocytes (red blood cells). Red blood cells are released into circulation as reticulocytes, then lose their nucleus fragments to become mature erythrocytes, which are a biconcave disc. Myeloid stem cells are also known as Colony Forming Units-GEMM, and directly differentiate into basophils and eosinophils, but must become promeylocytes before changing into neutrophils and macrophages/monocytes, and take several steps before finally turning into erythrocytes. Hormones and the microenvironment, i.e. the presence or absence of certain cytokines (interleukin 3, stem cell factor) and growth factors (granulocyte colony stimulating factor) regulate blood cell production.

The formation of erythrocytes specifically is called erythropoiesis. It requires erythropoietin, a cytokine produced in the kidneys that helps induce erythrocyte formation from CFU-GEMM cells. Erythrocytes need vitamins B9 and B12, iron and other minerals to develop. During development, they squeeze out their nucleus, which is eaten by nearby macrophages. They are the most numerous type of blood cell, and live for 120 days until they are broken down in the spleen and liver at a rate of 3 million per second. Most of the iron in their haemoglobin is recycled, but the rest of the haeme is degraded and excreted. Iron is necessary for erythrocyte development as it is part of haemoglobin. B12 and B9 are cofactors in bone marrow for stem cell growth, B6 binds to haemoglobin to increase oxygen binding affinity, and B1 (as TTP) is needed for DNA repair. It is the reticulo-endothelial system (in the spleen and liver) that removes old and damaged erythrocytes. Haemoglobin has divalent iron and globin chains, which are recycled, and a porphyrin ring (haeme), which is eliminated. Iron goes back to the bone marrow to form new erythrocytes, the amino acids from globin chains are used for protein synthesis and energy, and the porphyrin ring is broken into carbon monoxide and biliverdin (which becomes bilirubin for excretion). Some diseases related to erythropoiesis are porphyria, an issue with haeme production; iron deficiency anaemia; haemochromatosis (too much iron); sickle cell disease and thalassaemia, which are problems with globin formation.

There are far less white blood cells (leukocytes) than erythrocytes, with the ratio between the two at 1:700. Unlike erythrocytes, they have a nucleus, and are responsible for protecting the body from infections and cancer. Lymphocytes and monocytes/macrophages have a relatively clear cytoplasm, but neutrophils, basophils and eosinophils are filled with granules. B cell lymphocytes are responsible for antibody production, while T cells kill virus-infected and cancer cells (cytotoxic) recruit other types of leukocytes (inflammatory) and assist B cells in antibody production (helper), depending on their type. Some, known as suppressor T cells, inhibit T cell proliferation to decrease the immune response. Lymphocytes require many nutrients for their growth and function. Vitamins B9 and B12 act as cofactors in the bone marrow for stem cell growth, while vitamin C and alanine stimulate lymphocyte production. Glutamine also stimulates lymphocyte growth, partly because it is an important food source; glutathione assists in transporting nutrients to lymphocytes and beta-carotene and lycopene protect the lymphocytes from free radicals. DMG (dimethylglycine) and vitamin A stimulate lymphocyte production and proliferation in response to antigens from pathogens, while vitamin E enhances the lymphocyte's functioning. B cells stay in the bone marrow as they mature, unlike the T cells which mature in the thymus gland. They respond to the antigens on pathogens by transforming and producing antibodies; and make up less than 25% of all lymphocytes. Antibodies are a large group of proteins called globulins. These are made of 4 polypeptide chains, 2 identical heavy chains and 2 identical light chains joined by disulfide bonds. They are produced in lymph nodes and the spleen by B cells, after memory T cells instruct them to make the right antibodies. Memory T cells do this by storing a "snapshot" of antigens from every pathogen that has invaded the patient from a specific time frame.

Monocytes are the precursors to macrophages, and also produce interleukin-3 (IL-3). Like other blood cells, they require vitamins B9 and B12 as cofactors for the growth of stem cells. Glutamine and vitamin E enhance their ability to perform phagocytosis (eating pathogens and other unwanted substances), while beta-carotene improves their ability to present antigens to T cells. Some types of macrophages that monocytes develop into are tissue macrophages, osteoclasts and Kupffer cells. Macrophages are large leukocytes that also engulf microbes and debris from dead cells, even *Candida albicans*, as well as present antigens to lymphocytes. They can produce platelet-derived growth factor too, which promotes healing and the remodelling of tissues. Macrophages promote blood clotting by secreting thromboplastin, and mediate the acute phase response with interleukin 1 and tumour necrosis factor. Apart from vitamins B9 and B12 for stem cell development, vitamin B5 is necessary for the production of macrophages and glutamine is an important fuel source. Macrophages responsible for capillary and fibroblast growth need to be activated by glucans, which are found in oats. Both acetyl-l-carnitine and coenzyme Q10 enhance macrophage function, while vitamin D stimulates activity. Vitamin C is also necessary for macrophages, so they can produce interferons. Interferons are glycoproteins that destroy some types of harmful bacteria and block reproduction of some cancer cells and viruses.

Neutrophils are the most abundant of the leukocytes. These squeeze through the capillary walls into infected tissues so they can kill and engulf invading pathogens, such as bacteria. This is happening constantly, because despite the fact that most bacteria inside the human body are considered "safe", they are only safe because the neutrophils are keeping their numbers under control. Glutamine, acetyl-l-carnitine, taurine, and vitamins A, B6, C and E are all required for neutrophils to function.

The cytotoxic eosinophils are usually found at low levels, except in certain infections such as those of parasitic worms. They are also associated with allergic reactions. Little is known of what factors influence production and activity, but T cells may have some control over them and they are severely impaired by corticosteroids. Basophils are a type of mast cell that travels to sites of infection and inflammation in order to increase blood flow and inflammation. They work by releasing the contents of their granules, which contain histamine, serotonin, prostaglandins and leukotrienes. These chemicals are involved in some allergic reactions, including hay fever and anaphylactic reactions. In order to function, basophils require iron to help IgE antibodies bind to it for communication and quercetin to prevent the excessive release of histamine. Mast cells, besides basophils, are found in the connective tissue. They release heparin to inhibit inappropriate blood clotting and mediate inflammation, as they release histamine and serotonin when IgE antibodies bind to them. Quercetin stabilises the cell membranes of mast cells, and tyramine stimulates the release of histamine while zinc inhibits it.

7: Liver Detoxification

The liver performs over 200 functions, including detoxification, nutrient storage, breakdown of old erythrocytes, bile production and secretion, protein synthesis and metabolism, lipid metabolism and cholesterol synthesis.

Around 2 litres of blood is filtered by the liver every minute. This is vital because blood from the intestines contains high levels of bacteria and their endotoxins, antigen-antibody complexes and other substances. 99% of all bacteria and toxins (unless they are absorbed through the skin or lungs, or are deposited into tissues by chylomicrons before these reach the liver) are cleared during the first pass. Most metabolised compounds are then excreted through the bile and urine, though some exit the body through other routes such as the saliva. The liver synthesises and secretes around 1 litre of bile every day. Bile carries toxins for eventual excretion in the colon, but first it is stored in the gall bladder until a person eats, so the bile can assist in fat digestion. The bile breaks large globules of fat into smaller ones so there is more surface area for the enzymes to "attack", thereby speeding fat digestion. Later in the colon, residual bile and toxins are absorbed by fibre and excreted.

There are 2 phases of liver detoxification. Phase one detoxification transforms toxins into intermediate metabolites so they can be processed by phase two, which conjugates IMs with other compounds to make them even more water soluble or so they can be eliminated through the bile. There are 6 phase two pathways: glutathione conjugation, amino acid conjugation, methylation, sulphation, acetylation and glucoronidation. Phase one involves 50-100 enzymes all called cytochrome P450, which detoxify specific chemicals. Cytochrome P450 (CYP450) enzymes make toxins more water soluble so they can be immediately excreted by the kidneys or it will turn them into more active (dangerous) forms to be metabolised in phase two. Levels of CYP450 enzymes are affected by genetics, nutritional status and toxin exposure. The reactions performed by these enzymes are oxidation, reduction, hydrolysis, hydration and dehalogenation depending on the toxin and what enzyme is working on it. Toxins that are very lipid soluble will often be stored in adipose (fat) tissue, so weight loss can increase toxic load on the body. Both toxins and nutrients can speed up or inhibit phase one detoxification. Alcohol, caffeine, nicotine, dioxin, organophosphorous pesticides, paint and exhaust fumes, sulfonamides and barbiturates increase the rate of phase one; nutritional deficiencies, bacterial endotoxins, antihistamines, benzodiazepines and cimetidine inhibit it. Although they are safe and beneficial, naringenin from grapefruit, curcumin in turmeric, capsaicin in hot chilli, eugenol in clove oil and the herb *Calendula officinalis* all inhibit phase one too. The primary nutrients for phase one detoxification are vitamin C, copper, magnesium and zinc, while the secondary nutrients are iron, branched chain amino acids (isoleucine, leucine and valine), and vitamins B1, B2, B3, B6, B9 and B12. Overactive phase one detoxification results in elevated free radicals, but underactive phase one causes intolerances to caffeine, perfume and various environmental chemicals. Many drug interactions involve CYP450 enzymes being either induced or inhibited. The most important of these are CYP1A2, CYP2D6 and CYP3A4, which metabolise over half of most drugs.

Phase two detoxification must also be supported, because it protects the body against carcinogenic and generally dangerous intermediate metabolites. Sulphation is one pathway, where IMs are conjugated with sulphur containing compounds. These include IMs from the drug acetaminophen, hydroxylamines and cortisol. Cofactors required by this pathway are the vitamin B6, molybdenum, cysteine, glutathione, methionine and taurine. Glucoronidation is the conjugation of IMs with glucoronic acid, which requires the assistance of gum-containing foods. Morphine, benzodiazepines, salicylates and phenols are some substances dealt with by glucoronidation. Glutathione conjugation is the primary phase two pathway, and finishes eliminating chemicals such as penicillin, styrene, toxic metals, petroleum distillates and bacterial toxins. Cysteine, glycine and glutamine are components of glutathione, while the nutrients that assist this pathway are methionine, selenium, zinc, and vitamins B2, B6 and C. Acetylation is another phase two pathway, which conjugates IMs with acetyl CoA. Caffeine, serotonin and sulfonamides are some of the substances eliminated by acetylation. Lipoic acid, and vitamins B1, B2, B3 and B5 are required to form acetyl CoA, while vitamin C is needed for acetylation to occur. Chemicals dealt with by amino acid conjugation include salicylates, nicotinic acid and aliphatic amines. This pathway needs glycine, glutamine, taurine, arginine and ornithine, because these amino acids are what is conjugated to the IMs. Methylation is the conjugation of IMs with methyl groups, such as heavy metals, dopamine and histamine. Methionine, choline, SAMe, vitamin B12 and folate are necessary for methylation, and di- and trimethyl glycine provide substrates for methylation. Besides all of these nutrients, the rate of glutathione conjugation is increased by limonene containing foods (e.g. citrus peel) and foods from the *Brassicaceae* family such as cabbage and broccoli. Glucoronidation is sped up by fish oils, cigarettes, the oral contraceptive pill, Phenobarbital and limonene containing foods. Additionally, apart from various nutritional deficiencies, sulphation is inhibited by NSAIDs (e.g. aspirin) and the yellow food dye tartrazine, while glucoronidation is slowed down by aspirin and probenecid.

Methylation is not just a pathway in phase two detoxification. This is also necessary for neurotransmitter synthesis, DNA repair, cardiovascular and bone health. The carrier of methyl groups for this pathway is S-adenosyl methionine (SAMe), which becomes homocysteine once the methyl group is donated. Homocysteine is likely to be toxic to many cells, and elevated levels are associated with an increased risk of cardiovascular disease (such as atherosclerosis), venous thrombosis, stroke, kidney stones, osteoporosis, Alzheimer's disease and other dementias, depression, multiple sclerosis, mental retardation and psoriasis. It also speeds up the aging process. This is because high homocysteine increases oxidative stress, induces thrombosis and impairs appropriate collagen cross-linking. Smoking, caffeine consumption, type 2 diabetes mellitus, hypothyroidism and drugs including methotrexate, nitrous oxide and azaribine are associated with high homocysteine. Substances that lower homocysteine levels are betaine, vitamins B12, B9, B6, B2 and B3, glycine, serine and choline.

A well-known function of the liver is the detoxification of alcohol. Alcoholics are known to have higher levels of acetaldehyde in the brain, which is an intermediate between alcohol and acetate and gives rise to a morphine-like substance. This may be because alcohol inhibits vitamin B3 absorption, which is needed by aldehyde dehydrogenase to convert acetaldehyde to acetate. Alcohol also inhibits the enzyme delta-6-desaturase, which converts linoleic acid to gamma linolenic acid. However, GLA has been shown to reduce alcohol cravings. Excessive alcohol consumption can lead to bacterial overgrowth in the intestines, which causes an increased toxic load, thereby damaging the liver. This often results in alcoholic liver disease, autoimmune diseases, irritable bowel, inflammation and malnutrition. For alcohol detoxification, B vitamins are recommended to correct the deficiencies, and choline, selenium and vitamins A, C and E are recommended because of their antioxidant abilities. In addition, magnesium assists in phase one detoxification; zinc assists the enzymes alcohol and aldehyde dehydrogenase in alcohol elimination; glutamine relieves cravings; taurine and glutathione assist in liver detoxification; tryptophan elevates mood and eicosapentaenoic acid (EPA) stabilizes cell membranes. Carnitine is also recommended because it is capable of reversing alcohol-induced fatty liver disease, and lipoic acid can help the liver regenerate too. Apart from the nutrients described above, improving digestion with apple cider vinegar before meals or pre/probiotics is recommended.

Fatty liver disease is a build-up of fat in the liver cells, which can damage the liver and lead to cirrhosis (scarring). Risk factors for this include obesity, diabetes and high blood triglyceride levels. Causes of liver cirrhosis include excessive alcohol consumption, some drugs and other chemicals, certain infections, chronic or autoimmune hepatitis and haemochromatosis. Cirrhosis is still considered irreversible, which is why it is important to treat liver diseases by correcting malnutrition, preventing further liver cell damage and improving regeneration.

8: Inflammation and Immunity

Inflammation, immune function and other functions of the body are partly regulated by fatty acids. Two fatty acids, linoleic acid and alpha linolenic acid, are essential, meaning they must be obtained from food. Linoleic acid is an omega-6 fatty acid with 18 carbons and two double bonds, while alpha linolenic acid is an omega-3 fat with 18 carbons and 3 double bonds. The term "omega" followed by a number refers to the first carbon with a double bond from the methyl (omega) end of a fatty acid. Derivatives of essential fatty acids produce eicosanoids, which modulate heart rate, blood pressure, blood clotting and immune function, among other processes. They are also neuroprotective and a necessary component of the central nervous system.

The exact precursors to eicosanoids are arachidonic acid, dihomo-gamma-linolenic acid (DGLA) and eicoapentaenoic acid (EPA). Eicosanoids include prostaglandins, leukotrienes and thromboxanes, and are produced by almost every cell in the body. They do not travel through the systemic circulation; instead they are local hormones, affecting the cells that produce them or the cells nearby. The processes eicosanoids directly influence are inflammation, smooth (visceral) muscle contraction and water and sodium excretion (which they increase). They also act as modulators for various processes, so some eicosanoids may stimulate a particular pathway while others will inhibit it. Eicosanoids are derived from polyunsaturated fatty acids that have been esterified into phospholipids so they could become part of the cell membrane. Many have a very short half-life, and are often excreted or inactivated after only a few minutes. Most research suggests that, on average, 16% of alpha linolenic acid ("omega-3") is converted into EPA, some of which is changed again to DHA. Both of these can be used to produce the anti-inflammatory eicosanoids, and this conversion may be enhanced by a diet low in saturated and polyunsaturated omega-6 fatty acids.

One product of the eicosanoids pathway is thromboxane A_2. Specifically, it is produced down the cyclo-oxygenase (COX) pathway by platelets. The functions of thromboxane A_2 are stimulating angiogenesis (formation of new blood vessels) and platelet aggregation, as well as blood clotting, vasoconstriction and bronchoconstriction. Prostacyclin PGI_2 is another product of the COX eicosanoids pathway. It is made by endothelial cells in the blood vessels and produces the opposite effects of thromboxane A_2, such as thrombolysis (breakdown of blood clots) as opposed to thrombosis. The COX pathway also leads to prostaglandin $E_2F_2D_2$, which increases vasodilation (E), cAMP (E), vasoconstriction (F), bronchoconstriction and smooth muscle contraction, while decreasing platelet aggregation, lymphocyte migration and interleukins 1 and 2. It seems to have a split personality. Yet another eicosanoid product is 5&12 hydroperoxyeicosatetraenoic acid (5&12 HETE). This is made in the lipo-oxygenase (LOX) pathway, to promote metastasis, prevent apoptosis, increase adhesive factors, stimulate vasodilation and stimulate proteolytic enzymes; therefore it seems to be cancer promoting. The leukotrienes LTC_4 and LTD_4 are also produced by the LOX pathway, this time in immune cells. They increase inflammation; increase vascular permeability so that other immune cells can travel to the site of inflammation easily; cause T cells to divide and aggregate; as well as stimulating bronchoconstriction.

These inflammatory molecules are derived from arachidonic acid (AA), which has several inhibitors/saboteurs. EPA hijacks the COX pathway from AA, while this would be converted to PGE_2 and thromboxanes 2, EPA becomes leukotrienes 3 and PGE_3, which are anti-inflammatory. In addition, both EPA and DHA inhibit the release of AA from cell membranes and prevent it from being incorporated into the phospholipid bilayer of the cell membrane in the first place, by, once again, hijacking its position. In addition, ginger inhibits production of cyclo-oxygenase, and vitamin E reduces AA synthesis from other polyunsaturated fatty acids.

Series 4 leukotrienes, mentioned above, are a product of AA and the LOX pathway. These are involved in abnormal blood clotting, and excessive production of them results in increased inflammation, leading to problems like asthma, allergies and psoriasis. In fact, they are 1,000-10,000 times more inflammatory than histamine. The herbs St Mary's thistle (*Silybum marianum*) and frankincense (*Boswellia serrata*) suppress the LOX enzyme that catalyses the production of series 4 leukotrienes; quercetin, glutathione and DGLA also inhibit them.

Inflammation has 3 stages. The first one is acute inflammation, which is the first response to tissue injury and is controlled by histamine, serotonin, bradykinin, prostaglandins and leukotrienes. Next is the immune response, when leukocytes are activated by antigens released from acute inflammation. It can be beneficial if there are antigens from pathogens and they are destroyed, or detrimental if it leads to chronic inflammation, the third stage. This involves interleukins 1-3, granulocyte-macrophage colony stimulating factor (GM-CSF), tumour necrosis factor alpha, interferons and platelet-derived growth factor. Interleukins, INF-gamma and tumour necrosis factor alpha are all cytokines, chemicals that initiate immune responses. TNF-alpha is cytotoxic to bacteria, tumours and healthy tissue, while others stimulate production of more cytokines and increase the proliferation, differentiation and activity of leukocytes.

During inflammation, blood vessels constrict and then dilate, leading to the heat and redness (erythema) found in inflammation. Increased permeability of these vessels also causes swelling of the affected area, and pain is caused by release of bradykinin and serotonin. IgG antibodies are also released in order to stimulate phagocytic leukocytes. Inflammation often leads to the presence of pus, comprised of dead tissue, monocytes and neutrophils. One problem with chronic inflammation is the production of collagenase and elastase, which break down collagen and elastin in connective tissue. However, there are various foods that help to reduce inflammation. Garlic, specifically the chemical allicin, inhibits the synthesis of prostaglandins, thromboxane and leukotrienes, and so does curcumin in turmeric (requires fat for absorption). Both pineapple (bromelain) and onions (quercetin) are COX-2 inhibitors, but pineapple also reduces bradykinin. Additionally, ginger lowers prostaglandin and leukotriene synthesis.

A serious consequence of chronic inflammation is the development of cancer, due to the actions of the inflammatory compounds and free radicals produced in response to tissue injury (or perceived injury). One chemical responsible is the transcription factor NF-kappa B, stimulated by bacteria, viruses, toxins, cytokines and oxidative stress. Besides regulating the synthesis of pro-inflammatory cytokines, it encourages cell proliferation and prevents apoptosis of damaged cells.

Recently, it has become known that adipose (fat) tissue produces pro-inflammatory cytokines, which contribute to obesity-related diseases such as metabolic syndrome, along with other inflammatory and autoimmune conditions. These are referred to as adipokines, and they affect vascular sclerotic processes, immunity, inflammation, food intake and insulin sensitivity. Adipocytes also synthesise some interleukins (1, 1-RA, 6, 8, 10) and TNF-alpha. Leptin is a well-known adipokine, which decreases food intake and upregulates (increases) energy use. It also increases phagocyte function and the release of free radicals by neutrophils, and may protect dendritic cells of the immune system from apoptosis. Adiponectin is another; this one is a powerful antiatherogenic factor because it stops monocytes from adhering to the blood vessel endothelium and prevents the synthesis of TNF and NF-kappa B.

However, in skeletal joints it is pro-inflammatory and is associated with rheumatoid arthritis and matrix degradation.

9: Amino Acid Biochemistry

Proteins in the body are constantly being synthesised and broken down, whether it is for structural or functional purposes, or to produce energy.

The Urea Cycle

The urea cycle take place in the liver, and is necessary in order to remove ammonia from the body, because high ammonia levels can lead to a coma. Ammonia is formed by deamination reactions in the body, the breakdown of amino acids by intestinal bacteria and in some foods. Watermelon is one food that supports the urea cycle, because it contains citrulline which supports ammonia detoxification.

Alanine is the second most prominent amino acid in circulation (the first is glutamine), and transfers nitrogen to the liver. Muscle cells produce alanine from pyruvate so it can transport the nitrogen. The liver then converts this nitrogen into urea for excretion, and turns the alanine back into pyruvate, which will be either used for energy or transformed to glucose for the muscles. Alanine can be produced from protein catabolism and by transamination of pyruvate by alanine transaminase, or from alpha-ketoglutarate by alanine aminotransferase (requires vitamin B6). The second method of alanine synthesis also turns pyruvate into the amino acid glutamate. Glutamate can be made from alpha-ketoglutarate too, and this process, which requires vitamin B6, also converts leucine to alpha-ketoisocaproate.

Cysteine and methionine are the two sulphur-containing amino acids, and their metabolism is linked to methylation. Methionine is converted to S-adenosyl methionine, which then is turned into S-adenosyl homocysteine. The homocysteine can be transformed back to methionine (needs choline or vitamins B12 and B9) or to cysteine, which needs vitamin B6. The cysteine pathway also requires the enzymes cystathionine synthase and cystathionase; the production of these can be impaired by genetic defects but some cases respond well to vitamin B6 therapy.

In addition, threonine is often converted to glycine with the threonine cleavage complex, and glycine and serine can be converted into each other with the enzyme serine hydroxymethyltransferase and vitamin B9. Serine can also be produced from 3-phosphoglycerate, which needs vitamin B3 and glutamate. Proline and ornithine can be converted to glutamate, using proline 5-carboxylate and glutamate semialdehyde as intermediate steps. Proline requires vitamins B2 and B3 to become glutamate. Phenylalanine is commonly converted to tyrosine by phenylalanine monooxygenase, which requires vitamin B3, vitamin C, vitamin B9 and iron. An absence of this enzyme due to a genetic mutation causes phenylketonuria; children with this condition must be on a restrictive diet to avoid mental retardation.

Purines and Pyrimidines

Purines are the nitrogenous bases adenine (AMP), guanine (GMP) and hypoxanthine. These are almost always broken down into uric acid during digestion, so they are not obtained directly through food. Purines are part of DNA, components of the cofactors FAD, NAD and CoA, used to store cellular energy (especially ATP), parts of signaling molecules and neurotransmitters.

Purines are mostly synthesised in the liver. First, ribose5-phosphate is converted to phosphoribosyl pyrophosphate (PPRP) and then to 5-phosphoribosylamine (PRPP), which requires glutamine. PRPP can then be made into purines and pyrimidines (cytosine, uracil, thymine). Glutamine is necessary because it donates its amide ring, but aspartate, glycine, formate and carbon dioxide can also be nitrogen donors. Magnesium, iron, molybdenum and vitamins B2 and B3 are important cofactors. The benefits of the uric acid that dietary purines are broken into are stimulating the central nervous system and its antioxidant abilities, but high levels of purines are a risk factor for gout. Pyrimidines are also the building blocks of DNA and RNA, and assist in other biochemical reactions such as glycogenesis. These are synthesised from glutamine, while thymine (DNA only) needs vitamin B1.

Pyruvate and Acetyl CoA

One of the amino acids pyruvate can be synthesised from is threonine, and this reaction depends on vitamins B2 and B3. The methylation pathway can also produce pyruvate from cysteine, but glutamate and serine are also required. Tryptophan, with alanine as an intermediate, and glycine, with serine as its intermediate, can be synthesised into pyruvate too. Then, pyruvate dehydrogenase can convert pyruvate into acetyl CoA. This reaction is irreversible, preventing acetyl CoA from turning back into pyruvate and undergoing gluconeogenesis. Acetyl CoA must be used for energy, lipogenesis, cholesterol synthesis or ketogenesis regardless of what it was produced from.

Substrates of the Kreb's Cycle

As stated in Chapter 4, various substrates in the Kreb's/TCA cycle can be synthesised from amino acids. Alpha-ketoglutarate can be made from glutamate by glutamate dehydrogenase, which requires vitamin B3 as NAD or NADP. An ammonium molecule is also produced. Another by-product of methylation is succinyl CoA, formed when cystathionine becomes cysteine and leaves behind alpha-ketobutyrate. Isoleucine and valine can also be used to produce succinyl CoA, and these reactions need vitamins B2, B3 and B12, but isoleucine needs biotin and magnesium too. Phenylalanine and tyrosine can be synthesised into fumarate, while pyruvate can be turned into oxaloacetate in a magnesium and biotin dependent reaction.

Neurotransmitters

There are over 50 different neurotransmitters in the body, including acetylcholine, serotonin and dopamine.

Serotonin is a monoamine that regulates mood and sleep, inhibits gastric secretions and stimulates smooth muscle. To synthesise serotonin, tryptophan is first converted to 5-hydroxy-tryptophan by tryptophan hydroxylase. This requires iron, calcium, magnesium and vitamins B3, B9 and C. Aromatic acid decarboxylase then turns this into 5-hydroxy-tryptamine, or serotonin. Vitamins B6 and C are needed for this, as are magnesium and zinc.

GABA is an inhibitory neurotransmitter that reduces nervousness, anxiety, irritability and concentration. First, alpha-ketoglutarate aminotransferase converts glutamine to glutamate, an excitatory neurotransmitter that increases short term memory. Magnesium and manganese are used as cofactors. Then, glutamate decarboxylase uses taurine and vitamin B6 to change this to GABA. Glutamate dehydrogenase and transaminase can perform the same reaction, this time with vitamin B3 (as NAD) and vitamin B6 respectively as cofactors. An ammonium molecule is cleaved off in the process.

Some other important neurotransmitters are catecholamines, including noradrenalin, adrenalin and dopamine. The former 2 are mostly involved with the fight/flight (sympathetic) response, while dopamine has an important role in mood regulation. The first reaction involved converts tyrosine to dihydroxyphenylalanine by the enzyme tyrosine hydroxylase, which requires vitamins B1, B3 and B6 as well as iron. Next, aromatic acid decarboxylase uses vitamin B6 and magnesium to turn this into dopamine. Some dopamine is transformed into noradrenalin by dopamine beta-hydroxylase in a copper and vitamin C dependent reaction. Finally, some of this is converted to adrenalin by SAMe, vitamin B6 and vitamin B12.

Acetylcholine is another catecholamine, but it is involved with the parasympathetic ("rest and digest" nervous system. It may be involved in consciousness, memory, cognition and motor control too. Depression and Parkinson's disease seem to feature high levels of this, while dementia and schizophrenia is associated with a deficiency. Acetylcholine is produced by the enzyme choline acetyltransferase, which needs vitamin B1 and magnesium, joining acetyl CoA and choline together.

10: Food–borne Disease

Food poisoning is a general term used to describe symptoms caused by toxins in foods. These toxins can be made by bacteria such as *Salmonella*, by the foods themselves e.g. poisonous fish and aflatoxin, or by contamination with toxins such as heavy metals. Factors that increase the risk of food poisoning are leaving food exposed to room temperatures for long periods, using the same cutting boards for uncooked meat and vegetables, using boards made from absorbent material e.g. plastic and some woods; and not properly washing utensils after using them on raw meat. Food that has been frozen and re-thawed multiple times and leftovers more than a day old also have a higher risk of being contaminated.

The symptoms of food poisoning vary by underlying cause. Common symptoms are nausea, vomiting, diarrhoea, gastroenteritis, fever, loss of appetite, stomach cramps and even convulsions. Food poisoning can be caused by bacteria, viruses, moulds, fungi, algae, protozoa, helminths (worms) and fish.

Pathogenic (disease causing) bacteria cause trouble by invading the intestinal mucosa, and can produce anything from mild to severe symptoms. *Salmonella typhi* is one species of bacteria that can cause severe symptoms, in this case typhoid fever, which leads to blood poisoning, jaundice and bloody stools. It is spread by water and direct contact with infected people. Other bacteria of the *Salmonella* genus are found in animal products, especially chicken, and only cause gastroenteritis (fever, diarrhoea, vomiting and cramps). *Vibrio parahaemolyticus* also causes gastroenteritis, and is associated with eating raw seafood. Infection with *Shigella* species results in dysentery (vomiting, diarrhoea, fever and abdominal cramps), which is spread by poor hygiene and is mostly carried by vegetables. *Vibro cholerae* is also spread by poor hygiene/contaminated water. It causes severe symptoms, such as nausea, vomiting, diarrhoea, dehydration, shock, acidosis, renal failure and even death. *E. coli* is usually harmless, but milk contaminated with some strains can cause infantile diarrhoea. *Listeria monocytogenes* also only affects babies and immunocompromised adults (including pregnant women), and is found in contaminated dairy products. It can survive temperatures of up to 42 degrees Celsius. *Staphylococcus aureus*, or golden staph, is another cause of food poisoning. Humans often carry it on their hands and in the nasal passages without issues, but it causes nausea, vomiting/retching, diarrhoea and abdominal cramps in the gastrointestinal tract. One very dangerous species of bacteria is *Clostridium botulinum*, which not only causes gastrointestinal symptoms but also neurological symptoms, from double vision to respiratory failure. Incorrectly canned food, very moist cheese and smoked/pickled meat and fish are possible sources of this species of bacteria. It can survive in acidic conditions and has a wide temperature range. Finally, *Bacillus cereus*, which can survive boiling and frying, has two different enterotoxins: one that causes nausea, vomiting/retching, diarrhoea and cramps, and one that does not cause nausea and vomiting. The one that does cause vomiting is only found in rice while the other is found in a wide variety of foods.

Gastroenteritis is caused by many viruses. One of the most common is the Norwalk or rotavirus. Symptoms of rotavirus infection are severe diarrhoea, nausea, vomiting and respiratory symptoms, which can last from 4 days up to weeks. It is transmitted by contaminated water and the faecal-oral route (i.e. people who don't wash their hands). Hepatitis A is transmitted in the same way, and has a 2 week incubation period followed by fever, weakness/lethargy and jaundice. The infected person is contagious both during the incubation period and for 5-7 days after the symptoms have disappeared.

Moulds and fungi also cause food poisoning. These are parasites that feed off of dead and decaying organisms. Some produce aflatoxin, which may cause liver and/or kidney disease and is found in grains and legumes (including peanuts) that were wrongly stored in moist places. Rickettsia is neither a mould nor fungi, and is even smaller than viruses. It causes flu-like symptoms, and is an intracellular pathogen that can be found in dairy products and poultry. Toxins from algae, that are stored in the liver and glands of shellfish, can cause paralytic shellfish poisoning, which paralyses the heart and lungs and is therefore very dangerous. Additionally, freshwater, or blue-green algae, produce the possibly carcinogenic cyclic polypeptides.

One example of food poisoning from protozoa is amoebic dysentery, from seafood. Either *Balantidium coli* or *Entamoeba histolytica* can be responsible, and cause diarrhoea, ulceration of the small intestine and even abscesses in the liver and brain because they penetrate the intestinal walls. *Giardia lamblia* is spread through contaminated water, and causes gastroenteritis. There is also *Toxoplasma gundii*, which causes toxoplasmosis, a disease featuring swollen lymph nodes, fever and a high lymphocyte count. It is transmitted through undercooked beef and lamb, as well as cat faeces. Helminths, such as tape worms, Trichinella, roundworms and anisakis, are transmitted by contaminated and undercooked meats.

Fish themselves, as opposed to pathogens contaminating them, are also a source of food poisoning due to toxins they produce themselves. One type of fish poisoning is ciguatera, which has both gastrointestinal and neurological symptoms such as vertigo and myalgia. Trevally, red emperor, red bass, mackerel, coral trout and sweetlip can all cause ciguatera poisoning. The liver, intestines, gonads and skin of the tetrodontiforms (puffer fish, toadfish and blow fish) contain toxins that cause tetrodotoxism. This is dangerous because its symptoms include paralysis, respiratory distress and even brain damage in some people. Sardines, herring and anchovies can cause clupeoid poisoning, which may result in severe gastrointestinal and neurological disturbances, and even cardiovascular collapse. Lastly, scombroid poisoning, from anchovies, mackerel, sardines, tuna and pilchards, causes symptoms similar to histamine intoxication. Scombroid can be fatal because of the cardiovascular collapse and bronchiolar constriction.

Despite the dangers of food poisoning, it can be prevented in many ways. One method of prevention is to wash hands, kitchen countertops and all utensils with hot, soapy water, and then dry everything with paper towels. When cooking, meat must be thoroughly cooked instead of being left partially pink, eggs must have a firm yolk and white, and fish should be opaque. Sauce, soup and gravy should be brought to a boil. As for storage, perishables must be eaten or refrigerated within 2 hours, and food must never be defrosted at room temperature. Fish poisoning can be avoided by removing the parts of the fish containing the toxins, buying them from sources known to be safe, or by not eating that species of fish. Additionally, green tea, cumin seeds, grapefruit seed extract, garlic and onions can kill some forms of disease-causing bacteria. In fact, grapefruit seed extract can inhibit the growth of around 770 types of bacteria. There are also several methods of processing food to prevent bacterial growth, including pasteurisation and sterilisation, which are two methods of heating. Irradiation is sometimes used to kill all microbes, and ultra-filtration removes microbes from beer, wine and clear fruit juices. Some packaging methods, such as canning and bottling, where food is cooked in the containers to sterilise them, protect it from contamination and microbe growth if done correctly.

References

o Atsushi, O, et al. 2013, 'Basophils are required for the induction of Th2 immunity to haptens and peptide antigens', *Nature Communications* 4, Article number: 1738, 2013/04/23/online. dx.doi.org/10.1038/ncomms2740

o Bell, SG, Vallee, BL, 2009,'The Metallothionein/Thionein System: An Oxidoreductive Metabolic Zinc Link', *ChemBioChem*, vol.10, pp.55-62

o Berra, et al, 2006, 'The hypoxia-inducible-factor hydroxylases bring fresh air into hypoxia signalling', *EMBO reports*, vol.7, pp.41 – 45, http://www.nature.com/embor/journal/v7/n1/fig_tab/7400598_F2.html, accessed 28th November, 2013

o Bianchi, G, et al., 2006, 'Update on nutritional supplementation with branched chain amino acids', *Clinical Nutrition and Metabolic Care*, vol.8, pp.83-87

o Bode, C & Bode, JC, 1997, 'Alcohol's role in gastrointestinal disorders', *Alcohol Health & Research World*, vol.21, no.1

o Bowen, R, 2007, *Absorption of Lipids*, http://www.vivo.colostate.edu/hbooks/pathphys/digestion/smallgut/absorb_lipids.html, accessed November 2013

o Brehm, BJ & D'Alessio, DA, 2008, 'Weight loss and metabolic benefits with diets of varying fat and carbohydrate content: separating the wheat from the chaff', *Nature Clinical Practice*, vol.4, no.3, pp.140-146,

Nutritional Biochemistry Explained

http://www.nature.com/nrendo/journal/v4/n3/pdf/ncpendmet0730.pdf, accessed 6th January, 2014

o Demiroren, K, et al., 2013, 'Protective effects of L-carnitine, N-acetylcysteine and genistein in an experimental model of liver fibrosis', *Clinics and Research in Hepatology and Gastroenterology,* vol.38, no.1, pp.63-72

o Depeint, F, et al, 2006, 'Mitochondrial function and toxicity: role of the B vitamin family on mitochondrial energy metabolism', *Chemico-Biological Interactions*, vol.163, pp.94-113

o Fargarasan, S & Honjo, T, 2004, 'Regulation of IgA synthesis at mucosal surfaces', *Current Opinion in Immunology*, vol.16, pp.277–283

o Gilbert's Syndrome, n.d., *Gilberts Syndrome*, http://www.gilbertssyndrome.com/detoxification.php, accessed 8th January 2014

o Goljan, EF, 2014, *Rapid Review Pathology*, 4th edn. Elsevier

o Gordon, 2005, *Food and Nutrient Structure (in Human Nutrition)*, 11th edn. Geissler and Powers

o Gropper, SS, Smith, JL, 2013, *Advanced Nutrition for Human Metabolism,* 6th Edition, Wadsworth Publishing, USA

o Hoey, L et al 2009, 'Studies of biomarker responses to intervention with vitamin B12: a systematic review of randomised controlled trials', *American Journal of Clinical Nutrition*, 89, Supplement 1S-16S

Alexandra Preston

o Horton, ES, 1989, 'Metabolic fuels, utilization and exercise', *The American Journal of Clinical Nutrition*, vol.49, pp.931-937

o Hurrell, RF, et al, 1999, 'Inhibition of non-haem iron absorption in man by polyphenolic-containing beverages', *British Journal of Nutrition*, vol.81, pp.289-295

o Kato, et al., 2013, 'Gut TFH and IgA: key players for regulation of bacterial communities and immune homeostasis', *Immunology and Cell Biology*, vol.92, pp.49-56

o Khoshbaten, M, et al., 2010, 'N-Acetylcysteine Improves Liver Function in Patients with Non-Alcoholic Fatty Liver Disease', *Hepatitis Monthly*, vol.10, no.1, pp.12-16

o Lee, KW & Lee, HJ, 2006, 'Biphasic effects of dietary antioxidants on oxidative stress-mediated carcinogenesis', *Mechanisms of ageing and development*, vol.127, pp.424-431

o Libby, P & Ridker, PM, 2006, 'Inflammation and Atherothrombosis, From Population Biology and Bench Research to Clinical Practice', *Journal of the American College of Cardiology*, vol.48, no. 9, SupplA

o Linus Pauling Institute, http://lpi.oregonstate.edu/

o McMurry, J, 1992, *Organic Chemistry,* 3rd Ed. Brooks/Cole Publishing Company California, USA

o Malaguarnera, M, et al., 'L -Carnitine Supplementation to Diet: A New Tool inTreatment of Nonalcoholic Steatohepatitis — A Randomized and Controlled Clinical Trial', *The American Journal of Gastroenterology*, vol.105, pp.1338–1345 o Marriage, B, et al, 2003, 'Nutritional cofactor treatment in mitochondrial disorders', *Journal of the American Dietetic Association*, vol.103, no.8, pp.1029-1038

o Marsche, G, et al, 2013, 'Inflammation alters HDL composition and function: Implications for HDL-raising therapies', *Pharmacology & Therapeutics* vol.137, pp.341–351

o Marks, DB, Marks, AD, Smith CM, 1996, *Basic Medical Biochemistry – A Clinical Approach*, Lippincott Williams & Wilkins, Maryland USA

o Michelazzo, FB, et al., 2013, 'The Influence of Vitamin A Supplementation on Iron Status', *Nutrients*, vol.5, pp.4399-4413

o Mock, 2005, *Encyclopaedia of Human Nutrition*, Elsevier Ltd. USA

o National Digestive Diseases Information Clearinghouse, http://digestive.niddk.nih.gov/ddiseases/pubs/cirrhosis_ez/#causev, accessed 18th December, 2013

o Osiecki, H, 2002, *Cancer: A Biochemical & Nutritional Approach,* 2nd Edn. Bioconcepts Publishing, Queensland, Australia

o Patak, P, et al, 2004, 'Vitamin C is an important cofactor for both adrenal cortex and adrenal medulla', *Endocrine Research*, vol.30, no.4, pp.871-878

o Paulev-Zubieta, New Human Physiology, http://www.zuniv.net/physiology/book/chapter22.html, accessed 22nd November, 2013 o Pizzorno & Murray, 2006, *Homocysteine metabolism: nutritional modulation and impact on health and disease*

o Pizzorno & Murray, 2006, *Cytokines and Eicosanoids*, Chapter 42, pp.565-567

o Prosser, NR, et al, 2010, 'Influence of an iron intervention on the zinc status of young adult New Zealand women with mild iron deficiency', *The British Journal of Nutrition*, vol.104, pp.742-750

o Resta, SC, 2009, 'Effects of probiotics and commensals on intestinal epithelial physiology: implications for nutrient handling', *The Journal of Physiology*, vol.587, pp.4169-4174

o Roach, B, 2003, *Metabolism and Nutrition*, 2nd Edn. Mosby Int. London, UK.

o Russell, 2010, *iGenetics*, http://www.mun.ca/biology/scarr/iGen3_02-08.html

o Saxelby, CM, 2002, 'Dietary Implication. Advising patients about fat in the diet', *MJA*, vol.176, s123-s124

o Schachter, J, et al., 2005, *Effects of Omega 3 fatty acids on mental health. Evidence Report/technology assessment no. 116.* AHRQ publication no.05-E022-2, Rockville

o Schoendorfer, N, Davies, PSW, 2012, *Micronutrients. Micronutrient Interrelationships, synergism and antagonism*

o Shils, ME, Olson, JA, Shike, M, Ross, AC, 2006, *Modern Nutrition in Health and Disease*, 9th Edn. Lippincott Williams & Wilkins, Maryland, USA

Nutritional Biochemistry Explained

o Stankovic, MN, et al., 2013, 'The Effects of a-Lipoic Acid on Liver Oxidative Stress and Free Fatty Acid Composition in Methionine–Choline Deficient Diet-Induced NAFLD', *Journal of Medicinal Food,* vol.17, no.2, 254-61

o Touro University, 2013, *Glycation and diabetes mellitus,* http://research.tu.edu/laboratories/gugliucci/researchtopics/glycatio n.html, accessed 24th December, 2013

o US Department of Health and Human Services, 2012, *Toxicology of Chromium,* http://www.atsdr.cdc.gov/toxprofiles/tp7.pdf

o Varela-Moreiras, GV, et al, 2009, 'Cobalamin, folic acid and homocysteine', *Nutrition Reviews,* vol.67, no.1, pp.569-572

o Waring, et al, (unpublished), 'Report on Absorption of magnesium sulfate (Epsom salts) across the skin', School of Biosciences, University of Birmingham, United Kingdom, http://www.epsomsaltcouncil.org/articles/report_on_absorption_of _magnesium_sulfat e.pdf, accessed 15th December 2013

o Watanabe, F, 2013, 'Biologically Active Vitamin B12 Compounds in Foods for Preventing Deficiency among Vegetarians and Elderly Subjects', *Journal of Agriculture and Food Chemistry,* vol.61, pp.6769–6775

o Watermelon benefits, *Urea Cycle,* http://watermelonbenefits.com/citrulline/, accessed 23rd December 2013

o Web Ics Purdue Edu, *Nutrient absorption,* http://web.ics.purdue.edu/~smills/ANSC230/Digestive%20Physiol ogy/Absorption.html, accessed 18th December, 2013 o Westerterp, KR, 2009, 'Dietary fat oxidation as a function of body fat', *Current Opinion in Lipidology*, vol.20, pp.45-49

o Westman, EC, et al, 2007, 'Low-carbohydrate nutrition and metabolism', *The American Journal of Clinical Nutrition*, vol.86, pp.276-284

o Ye, Z and Song, H, 2008, 'Antioxidant vitamins intake and the risk of coronary heart disease: meta-analysis of cohort studies', *European Journal of Cardiovascular Prevention and Rehabilitation*, vol.18, no.1, pp.27-34

Yousuf, O, et al, 2013, 'High-Sensitivity C-Reactive Protein and Cardiovascular Disease, A Resolute Belief or an Elusive Link?', *JACC*, vol.62, no.5, pp.397–408

o Zhang, H, et al., 2010, 'Emerging role of adipokines as mediators in atherosclerosis', *World Journal of Cardiology*, vol.2, no.11, pp.370-376